Wireless Communication Networks and Systems

Wireless Communication Networks and Systems

Tom Sparks

STATES
ACADEMIC PRESS
www.statesacademicpress.com

States Academic Press,
109 South 5th Street,
Brooklyn, NY 11249, USA

ISBN: 978-1-63989-570-0

Cataloging-in-Publication Data

Wireless communication networks and systems / Tom Sparks.
p. cm.
Includes bibliographical references and index.
ISBN 978-1-63989-570-0
1. Wireless communication systems. 2. Telecommunication systems. I. Sparks, Tom.
TK5103.2 .W57 2022
621.384--dc23

For information on all States Academic Press publications
visit our website at www.statesacademicpress.com

STATES
ACADEMIC PRESS

Contents

Preface

Computer networks which use wireless data connections for transfer of information between two network nodes are known as wireless networks. Most of the wireless networks make use of radio waves for transferring data between different nodes of a network. Some of the other technologies employed for wireless communication are terrestrial microwaves, free space optical communication and satellites. These networks can be broadly categorized into wireless personal area networks (WPAN), wireless local area networks (WLAN), wireless ad hoc networks (WANET), cellular networks, global networks and space networks. A few examples of cellular networks are global systems for mobile communication, personal communication service and digital advanced mobile phone service. This book unfolds the innovative aspects of wireless communication networks and systems which will be crucial for the holistic understanding of the subject matter. It is an essential guide for both academicians and those who wish to pursue this discipline further. Coherent flow of topics, student-friendly language and extensive use of examples make this textbook an invaluable source of knowledge.

A short introduction to every chapter is written below to provide an overview of the content of the book:

Chapter 1 - The method of transmitting information from one point to other without using any physical medium such as wires or cables is known as wireless communication. The most common wireless technologies use radiowaves. This chapter introduces the reader to the key concepts related to wireless communication; **Chapter 2** - The medium through which a message is sent to its intended receiver is known as a communication channel. There are several types of communication channels such as digital channel, multiplexed channel, virtual channel, return channel, upstream and downstream channel, broadcast, unicast, and multicast channel, etc. The aim of this chapter is to explore the various types and models of communication channels; **Chapter 3** - The computer networks which are not connected by cables or wires are called wireless networks. There are various types of wireless networks such as wireless personal area network (PAN), wireless local area network (LAN), wireless ad hoc network, wireless metropolitian network (MAN), wireless wide area network (WAN), cellular network and global area network, etc. This chapter has been carefully written to provide an easy understanding of the varied facets of a wireless network; **Chapter 4** - Mobile telephony is the provision of telephone services to phones which can be moved from one place to another. Bluetooth is also a wireless technology which is used to transfer data between two devices over short distances using ultra high frequency radiowaves and building personal area network. All the diverse aspects of wireless communications have been carefully analyzed in this chapter; **Chapter 5** - Radio is a technology to send electromagnetic signals to deliver information from one place to another. Radio propagation is the behavior of radiowaves when they are transmitted from one point to another. The chapter closely examines the key concepts of radios to provide an extensive understanding of the subject; **Chapter 6** - The prevention of unauthorized damage to computers which are connected to wireless networks is known as wireless security. The process of monitoring unauthorized access points is known as intrusion detection, and the device that performs intrusion detection and takes countermeasures is known as the wireless intrusion prevention system. This chapter discusses in detail the theories and concepts related to wireless security.

Finally, I would like to thank my fellow scholars who gave constructive feedback and my family members who supported me at every step.

Tom Sparks

1

Wireless Communication: An Introduction

The method of transmitting information from one point to other without using any physical medium such as wires or cables is known as wireless communication. The most common wireless technologies use radiowaves. This chapter introduces the reader to the key concepts related to wireless communication.

Wireless

Wireless is a term used to describe telecommunications in which electromagnetic waves (rather than some form of wire) carry the signal over part or the entire communication path. Some monitoring devices, such as intrusion alarms, employ acoustic waves at frequencies above the range of human hearing; these are also sometimes classified as wireless.

The first wireless transmitters went on the air in the early 20th century using radiotelegraphy (Morse code). Later, as modulation made it possible to transmit voices and music via wireless, the medium came to be called "radio." With the advent of television, fax, data communication, and the effective use of a larger portion of the spectrum, the term "wireless" has been resurrected.

Common examples of wireless equipment in use today include:

- Cellular phones and pagers: Provide connectivity for portable and mobile applications, both personal and business.
- Global Positioning System (GPS): Allows drivers of cars and trucks, captains of boats and ships, and pilots of aircraft to ascertain their location anywhere on earth.
- Cordless computer peripherals: The cordless mouse is a common example; keyboards and printers can also be linked to a computer via wireless.
- Cordless telephone sets: These are limited-range devices, not to be confused with cell phones.
- Home-entertainment-system control boxes: The VCR control and the TV channel control are the most common examples; some hi-fi sound systems and FM broadcast receivers also use this technology.
- Remote garage-door openers: One of the oldest wireless devices in common use by consumers; usually operates at radio frequencies.
- Two-way radios: This includes Amateur and Citizens Radio Service, as well as business, marine, and military communications.

- Baby monitors: These devices are simplified radio transmitter/receiver units with limited range.
- Satellite television: Allows viewers in almost any location to select from hundreds of channels.
- Wireless LANs or local area networks: Provide flexibility and reliability for business computer users.

Wireless technology is rapidly evolving, and is playing an increasing role in the lives of people throughout the world. In addition, ever-larger numbers of people are relying on the technology directly or indirectly. (It has been suggested that wireless is overused in some situations, creating a social nuisance.) More specialized and exotic examples of wireless communications and control include:

- Global System for Mobile Communication (GSM): A digital mobile telephone system used in Europe and other parts of the world; the de facto wireless telephone standard in Europe.
- General Packet Radio Service (GPRS): A packet-based wireless communication service that provides continuous connection to the Internet for mobile phone and computer users.
- Enhanced Data GSM Environment (EDGE): A faster version of the Global System for Mobile (GSM) wireless service.
- Universal Mobile Telecommunications System (UMTS): A broadband, packet-based system offering a consistent set of services to mobile computer and phone users no matter where they are located in the world.
- Wireless Application Protocol (WAP): A set of communication protocols to standardize the way that wireless devices, such as cellular telephones and radio transceivers, can be used for Internet access.
- i-Mode: The world's first "smart phone" for Web browsing, first introduced in Japan; provides color and video over telephone sets.

Wireless can be divided into:

- Fixed wireless: The operation of wireless devices or systems in homes and offices, and in particular, equipment connected to the Internet via specialized modems.
- Mobile wireless: The use of wireless devices or systems aboard motorized, moving vehicles; examples include the automotive cell phone and PCS (personal communications services).
- Portable wireless: The operation of autonomous, battery-powered wireless devices or systems outside the office, home, or vehicle; examples include handheld cell phones and PCS units.
- IR wireless: The use of devices that convey data via IR (infrared) radiation; employed in certain limited-range communications and control systems.

Wireless Communication

Wireless Communication is the fastest growing and most vibrant technological areas in the communication field. Wireless Communication is a method of transmitting information from one point to other, without using any connection like wires, cables or any physical medium.

Generally, in a communication system, information is transmitted from transmitter to receiver that is placed over a limited distance. With the help of Wireless Communication, the transmitter and receiver can be placed anywhere between few meters (like a T.V. Remote Control) to few thousand kilometers (Satellite Communication).

We live in a World of communication and Wireless Communication, in particular is a key part of our lives. Some of the commonly used Wireless Communication Systems in our day – to – day life are: Mobile Phones, GPS Receivers, Remote Controls, Bluetooth Audio and Wi-Fi etc.

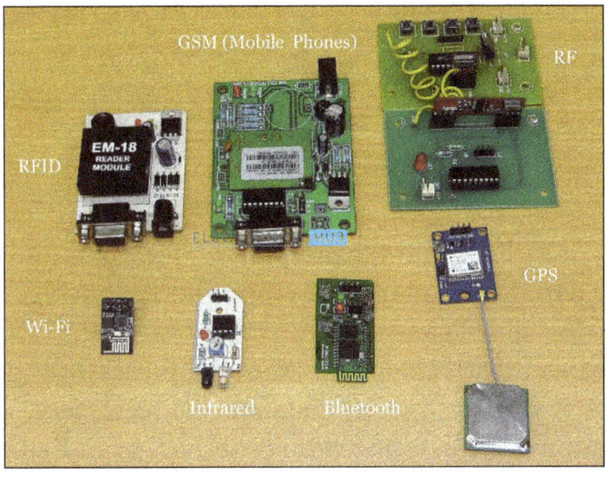

Communication Systems can be Wired or Wireless and the medium used for communication can be Guided or Unguided. In Wired Communication, the medium is a physical path like Co-axial Cables, Twisted Pair Cables and Optical Fiber Links etc. which guides the signal to propagate from one point to other. Such type of medium is called Guided Medium. On the other hand, Wireless Communication doesn't require any physical medium but propagates the signal through space. Since, space only allows for signal transmission without any guidance, the medium used in Wireless Communication is called Unguided Medium.

If there is no physical medium, then how does wireless communication transmit signals? Even though there are no cables used in wireless communication, the transmission and reception of signals is accomplished with Antennas.

Antennas are electrical devices that transform the electrical signals to radio signals in the form of Electromagnetic (EM) Waves and vice versa. These Electromagnetic Waves propagates through space. Hence, both transmitter and receiver consist of an antenna.

Electromagnetic Waves

Electromagnetic Waves carry the electromagnetic energy of electromagnetic field through space.

Electromagnetic Waves include Gamma Rays (γ – Rays), X – Rays, Ultraviolet Rays, Visible Light, Infrared Rays, Microwave Rays and Radio Waves. Electromagnetic Waves (usually Radio Waves) are used in wireless communication to carry the signals.

An Electromagnetic Wave consists of both electric and magnetic fields in the form of time varying sinusoidal waves. Both these fields are oscillating perpendicular to each other and the direction of propagation of the Electromagnetic Wave is again perpendicular to both these fields.

Mathematically, an Electromagnetic Wave can be described using Maxwell's equations. Pictorial representation of an Electromagnetic Wave is shown below where the Electric Field is acting in the Y – axis, magnetic field is acting in the Z – axis and the Electromagnetic Wave propagates in X – axis.

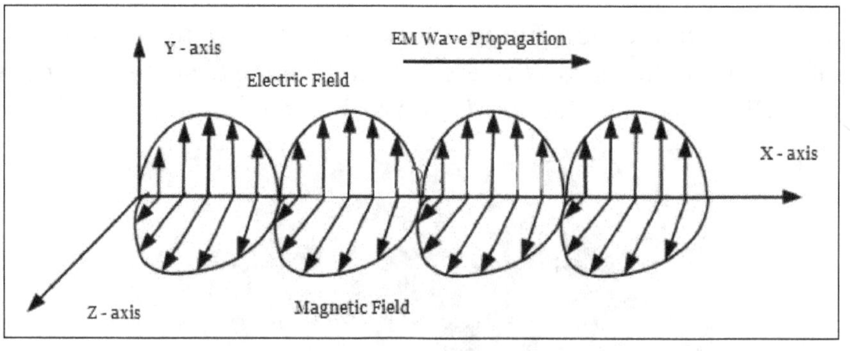

Since the use of smoke signals, flags and flashing mirrors in the pre – historic period, Wireless communication has been a part of human life and is continuously evolving. Modern Wireless Communication i.e. using electrical signals and radio waves for communication has been around us for more than 100 years. In the year 1897, Guglielmo Marconi successfully demonstrated the Wireless Telegraphy by sending EM Waves for a short distance of 100 meters. This demonstration paved way for Radio Communication and the term Radio is derived from Radiant Energy.

By early 1900's, Trans – Atlantic radio transmission had been established, where Marconi successfully transmitted messages in the form of Morse code. Since then, the technology related to wireless communication and wireless system has advanced rapidly and thus enabling transmissions over longer distances at low cost with cheaper devices.

Throughout the development of wireless communication, there are many wireless systems and methods that flourished and many got disappeared. The best example for this is Telephone Communication and Television Transmission. Initially, all telephone related communication was carried out (and still is) using wired network. But the rapid growth of Mobile Communication started to replace the complex wired telephone system. In this scenario, the wired technology became outdated and got replaced by wireless communication.

Another scenario where wireless communication got replaced by wired communication is Television broadcasting. In the early days, television signals were broadcasted using wireless radio transmitters. But this setup got replaced by cable television.

These two examples point out that with the development of technology; we always have to choose what's best for the situation i.e. in some areas we have to use wired communication whereas in the other, going for wireless may be a better option.

Need of Wireless Communication

When wired communication can do most of the tasks that a wireless communication can, why do we need Wireless Communication? The primary and important benefit of wireless communication is mobility. Apart from mobility, wireless communication also offers flexibility and ease of use, which makes it increasingly popular day – by – day. Wireless Communication like mobile telephony can be made anywhere and anytime with a considerably high throughput performance.

Another important point is infrastructure. The setup and installation of infrastructure for wired communication systems is an expensive and time consuming job. The infrastructure for wireless communication can be installed easily and low cost. In emergency situations and remote locations, where the setup of wired communication is difficult, wireless communication is a viable option.

Advantages of Wireless Communication

There are numerous advantages of Wireless Communication Technology, Wireless Networking and Wireless Systems over Wired Communication like Cost, Mobility, Ease of Installation, and Reliability etc.

- Cost: The cost of installing wires, cables and other infrastructure is eliminated in wireless communication and hence lowering the overall cost of the system compared to wired communication system. Installing wired network in building, digging up the Earth to lay the cables and running those wires across the streets is extremely difficult, costly and time consuming job.

  In historical buildings, drilling holes for cables is not a best idea as it destroys the integrity and importance of the building. Also, in older buildings with no dedicated lines for communication, wireless communication like Wi-Fi or Wireless LAN is the only option.

- Mobility: Mobility is the main advantage of wireless communication system. It offers the freedom to move around while still connected to network.

- Ease of Installation: The setup and installation of wireless communication network's equipment and infrastructure is very easy as we need not worry about the hassle of cables. Also, the time required to setup a wireless system like a Wi-Fi network for example, is very less when compared to setting up a full cabled network.

- Reliability: Since there are no cables and wires involved in wireless communication, there is no chance of communication failure due to damage of these cables which may be caused by environmental conditions, cable splice and natural diminution of metallic conductors.

- Disaster Recovery: In case of accidents due to fire, floods or other disasters, the loss of communication infrastructure in wireless communication system can be minimal.

Disadvantages of Wireless Communication

Even though wireless communication has a number of advantages over wired communication, there are a few disadvantages as well. The most concerning disadvantages are Interference, Security and Health.

- Interference: Wireless Communication systems use open space as the medium for transmitting signals. As a result, there is a huge chance that radio signals from one wireless communication system or network might interfere with other signals.

  The best example is Bluetooth and Wi-Fi (WLAN). Both these technologies use the 2.4GHz frequency for communication and when both of these devices are active at the same time, there is a chance of interference.

- Security: One of the main concerns of wireless communication is Security of the data. Since the signals are transmitted in open space, it is possible that an intruder can intercept the signals and copy sensitive information.

- Health Concerns: Continuous exposure to any type of radiation can be hazardous. Even though the levels of RF energy that can cause the damage are not accurately established, it is advised to avoid RF radiation to the maximum.

Basic Elements of a Wireless Communication System

A typical Wireless Communication System can be divided into three elements: the Transmitter, the Channel and the Receiver. Figure below shows the block diagram of wireless communication system.

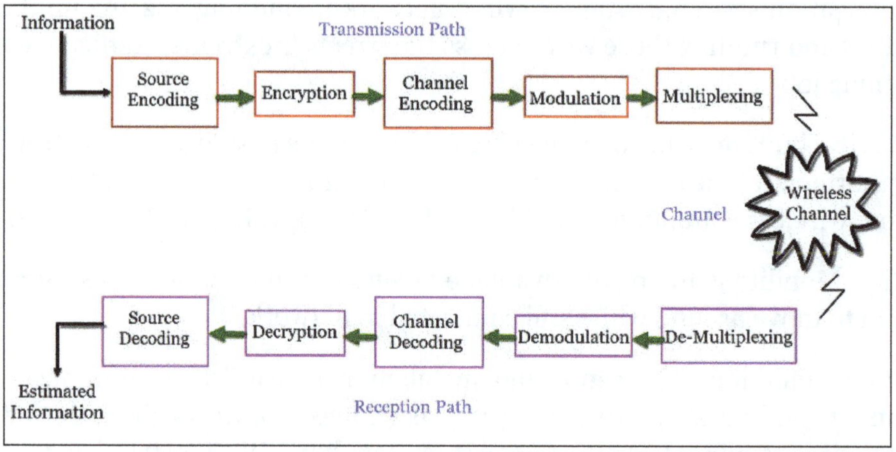

The Transmission Path

A typical transmission path of a Wireless Communication System consists of Encoder, Encryption, Modulation and Multiplexing. The signal from the source is passed through a Source Encoder, which converts the signal in to a suitable form for applying signal processing techniques. The redundant information from signal is removed in this process in order to maximize the utilization of resources. This signal is then encrypted using an Encryption Standard so that the signal and the information is secured and doesn't allow any unauthorized access.

Channel Encoding is a technique that is applied to the signal to reduce the impairments like noise, interference, etc. During this process, a small amount of redundancy is introduced to the signal so that it becomes robust against noise. Then the signal is modulated using a suitable Modulation Technique (like PSK, FSK and QPSK etc.), so that the signal can be easily transmitted using antenna.

The modulated signal is then multiplexed with other signals using different Multiplexing Techniques like Time Division Multiplexing (TDM) or Frequency Division Multiplexing (FDM) to share the valuable bandwidth.

The Channel

The channel in Wireless Communication indicates the medium of transmission of the signal i.e. open space. A wireless channel is unpredictable and also highly variable and random in nature. A channel maybe subject to interference, distortion, noise, scattering etc. and the result is that the received signal may be filled with errors.

The Reception Path

The job of the Receiver is to collect the signal from the channel and reproduce it as the source signal. The reception path of a Wireless Communication System comprises of Demultiplexing, Demodulation, Channel Decoding, Decryption and Source Decoding. From the components of the reception path it is clear that the task of the receiver is just the inverse to that of transmitter.

The signal from the channel is received by the Demultiplexer and is separated from other signals. The individual signals are demodulated using appropriate Demodulation Techniques and the original message signal is recovered. The redundant bits from the message are removed using the Channel Decoder. Since the message is encrypted, Decryption of the signal removes the security and turns it into simple sequence of bits. Finally, this signal is given to the Source Decoder to get back the original transmitted message or signal.

Types of Wireless Communication Systems

Today, people need Mobile Phones for many things like talking, internet, multimedia etc. All these services must be made available to the user on the go i.e. while the user is mobile. With the help of these wireless communication services, we can transfer voice, data, videos, images etc.

Wireless Communication Systems also provide different services like video conferencing, cellular telephone, paging, TV, Radio etc. Due to the need for variety of communication services, different types of Wireless Communication Systems are developed. Some of the important Wireless Communication Systems available today are:

- Television and Radio Broadcasting,
- Satellite Communication,
- Radar,
- Mobile Telephone System (Cellular Communication),
- Global Positioning System (GPS),
- Infrared Communication,

- WLAN (Wi-Fi),
- Bluetooth,
- Paging,
- Cordless Phones,
- Radio Frequency Identification (RFID).

There are many other system with each being useful for different applications. Wireless Communication systems can be again classified as Simplex, Half Duplex and Full Duplex. Simplex communication is one way communication. An example is Radio broadcast system.

Half Duplex is two way communications but not simultaneous one. An example is walkie – talkie (civilian band radio). Full Duplex is also two way communication and it is a simultaneous one. Best example for full duplex is mobile phones. The devices used for Wireless Communication may vary from one service to other and they may have different size, shape, data throughput and cost. The area covered by a Wireless Communication system is also an important factor. The wireless networks may be limited to a building, an office campus, a city, a small regional area (greater than a city) or might have global coverage.

- Television and Radio Broadcasting: Radio is considered to be the first wireless service to be broadcast. It is an example of a Simplex Communication System where the information is transmitted only in one direction and all the users receiving the same data.

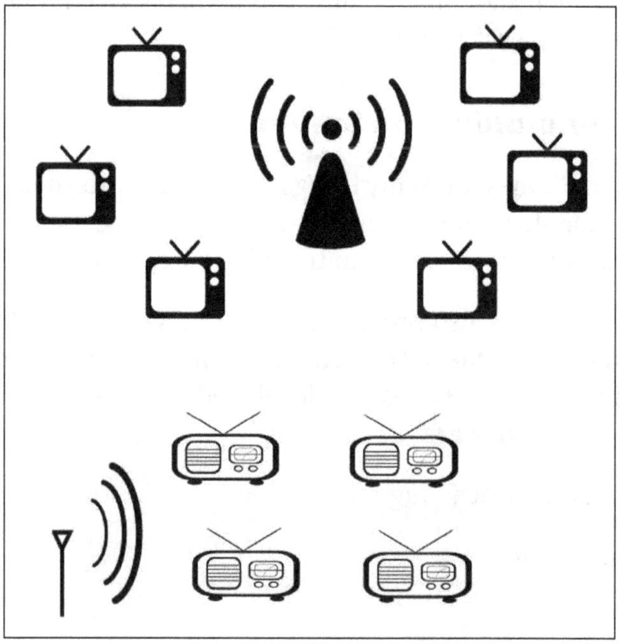

- Satellite Communication: Satellite Communication System is an important type of Wireless Communication. Satellite Communication Networks provide worldwide coverage independent to population density. Satellite Communication Systems offer telecommunication (Satellite Phones), positioning and navigation (GPS), broadcasting, internet, etc.

Other wireless services like mobile, television broadcasting and other radio systems are dependent of Satellite Communication Systems.

- Mobile Telephone Communication System: Perhaps, the most commonly used wireless communication system is the Mobile Phone Technology. The development of mobile cellular device changed the World like no other technology. Today's mobile phones are not limited to just making calls but are integrated with numerous other features like Bluetooth, Wi-Fi, GPS, and FM Radio.
- Global Positioning System (GPS): GPS is solely a subcategory of satellite communication. GPS provides different wireless services like navigation, positioning, location, speed etc. with the help of dedicated GPS receivers and satellites.
- Bluetooth: Bluetooth is another important low range wireless communication system. It provides data, voice and audio transmission with a transmission range of 10 meters. Almost all mobile phones, tablets and laptops are equipped with Bluetooth devices. They can be connected to wireless Bluetooth receivers, audio equipment, cameras etc.
- Paging: Although it is considered an obsolete technology, paging was a major success before the wide spread use of mobile phones. Paging provides information in the form of messages and it is a simplex system i.e. the user can only receive the messages.
- Wireless Local Area Network (WLAN): Wireless Local Area Network or WLAN (Wi-Fi) is an internet related wireless service. Using WLAN, different devices like laptops and mobile phones can connect to an access point and access internet.
- Infrared Communication: Infrared Communication is another commonly used wireless communication in our daily lives. It uses the infrared waves of the Electromagnetic (EM) spectrum. Infrared (IR) Communication is used in remote controls of Televisions, cars, audio equipment etc.

Wireless Transmission System

Wireless transmission is a form of unguided media. Wireless communication involves no physical link established between two or more devices, communicating wirelessly. Wireless signals are spread over in the air and are received and interpreted by appropriate antennas.

When an antenna is attached to electrical circuit of a computer or wireless device, it converts the digital data into wireless signals and spread all over within its frequency range. The receptor on the other end receives these signals and converts them back to digital data. A little part of electromagnetic spectrum can be used for wireless transmission.

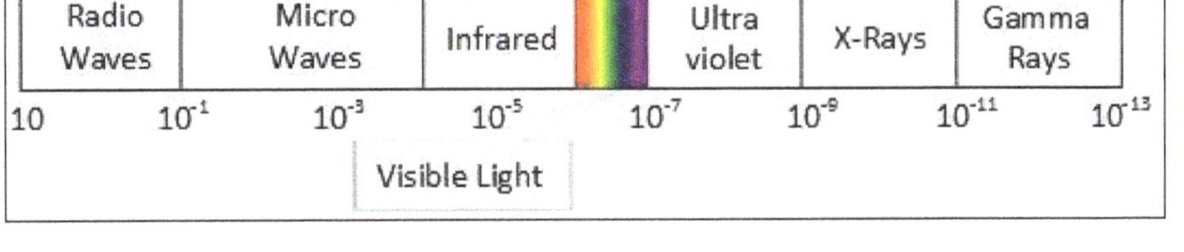

Radio Transmission

Radio frequency is easier to generate and because of its large wavelength it can penetrate through walls and structures alike. Radio waves can have wavelength from 1 mm – 100,000 km and have frequency ranging from 3 Hz (Extremely Low Frequency) to 300 GHz (Extremely High Frequency). Radio frequencies are sub-divided into six bands.

Radio waves at lower frequencies can travel through walls whereas higher RF can travel in straight line and bounce back. The power of low frequency waves decreases sharply as they cover long distance. High frequency radio waves have more power. Lower frequencies such as VLF, LF, MF bands can travel on the ground up to 1000 kilometers, over the earth's surface.

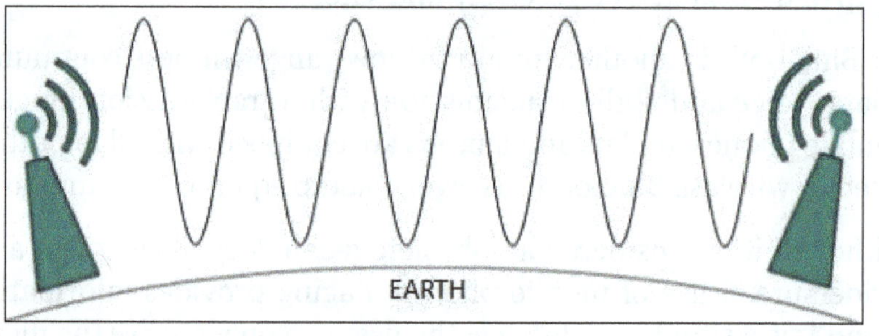

Radio waves of high frequencies are prone to be absorbed by rain and other obstacles. They use Ionosphere of earth atmosphere. High frequency radio waves such as HF and VHF bands are spread upwards. When they reach Ionosphere, they are refracted back to the earth.

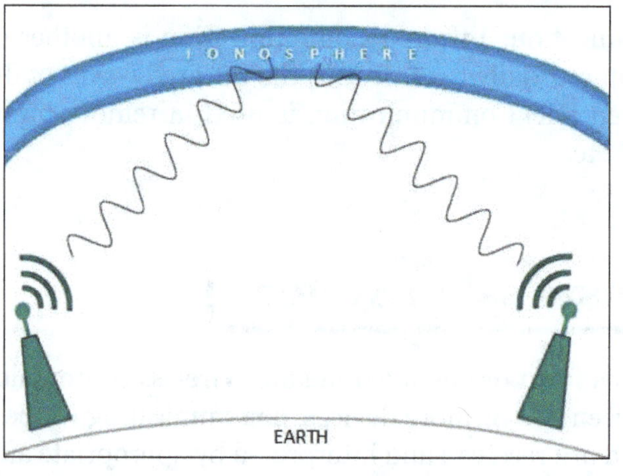

Microwave Transmission

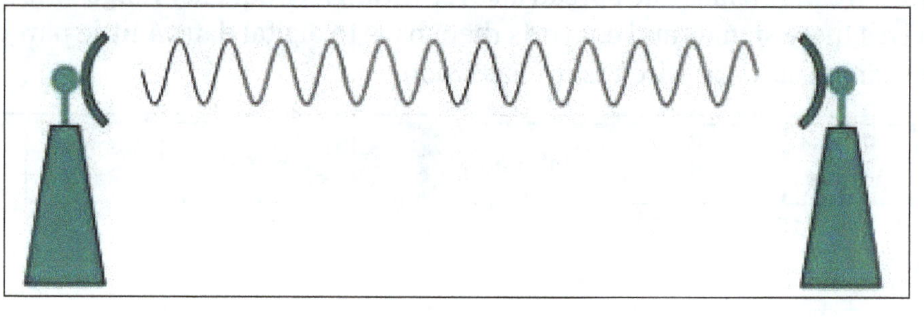

Electromagnetic waves above 100 MHz tend to travel in a straight line and signals over them can be sent by beaming those waves towards one particular station. Because Microwaves travels in straight lines, both sender and receiver must be aligned to be strictly in line-of-sight. Microwaves can have wavelength ranging from 1 mm – 1 meter and frequency ranging from 300 MHz to 300 GHz.

Microwave antennas concentrate the waves making a beam of it. Multiple antennas can be aligned to reach farther. Microwaves have higher frequencies and do not penetrate wall like obstacles. Microwave transmission depends highly upon the weather conditions and the frequency it is using.

Infrared Transmission

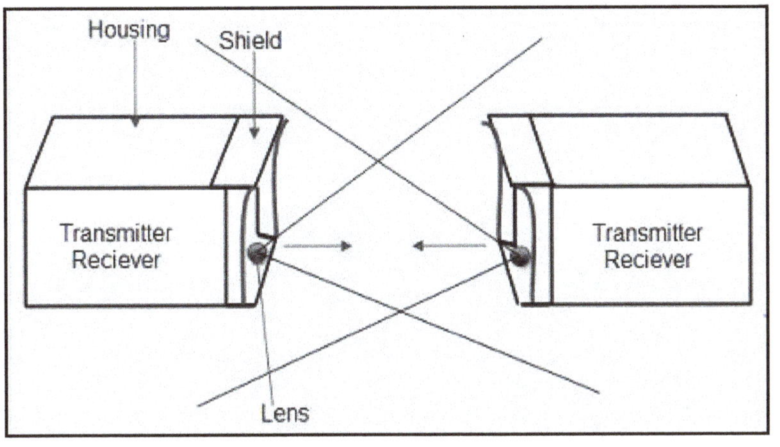

Infrared wave lies in between visible light spectrum and microwaves. It has wavelength of 700-nm to 1-mm and frequency ranges from 300-GHz to 430-THz. Infrared wave is used for very short range communication purposes such as television and it's remote. Infrared travels in a straight line hence it is directional by nature. Because of high frequency range, Infrared cannot cross wall-like obstacles.

Light Transmission

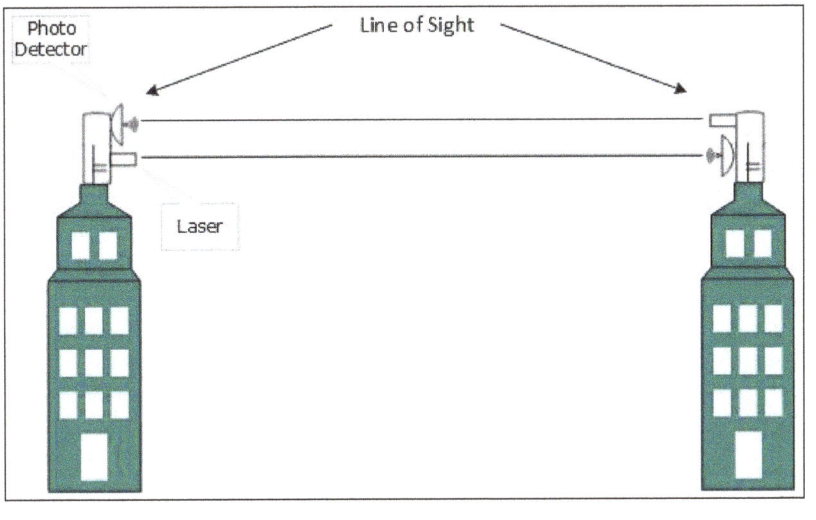

Highest most electromagnetic spectrum which can be used for data transmission is light or optical signaling. This is achieved by means of LASER. Because of frequency light uses, it tends to travel strictly in straight line. Hence the sender and receiver must be in the line-of-sight. Because laser transmission is unidirectional, at both ends of communication the laser and the photo-detector need to be installed. Laser beam is generally 1mm wide hence it is a work of precision to align two far receptors each pointing to lasers source.

Laser works as TX (transmitter) and photo-detectors works as Rx (receiver). Lasers cannot penetrate obstacles such as walls, rain, and thick fog. Additionally, laser beam is distorted by wind, atmosphere temperature, or variation in temperature in the path. Laser is safe for data transmission as it is very difficult to tap 1mm wide laser without interrupting the communication channel.

2

Communication Channel: Types and Models

The medium through which a message is sent to its intended receiver is known as a communication channel. There are several types of communication channels such as digital channel, multiplexed channel, virtual channel, return channel, upstream and downstream channel, broadcast, unicast, and multicast channel, etc. The aim of this chapter is to explore the various types and models of communication channels.

Communication Channel

A communication channel is a type of media that is used to transfer a message from one person to another. In business specifically, communication channels are the way information flows in the organization within, and with other companies.

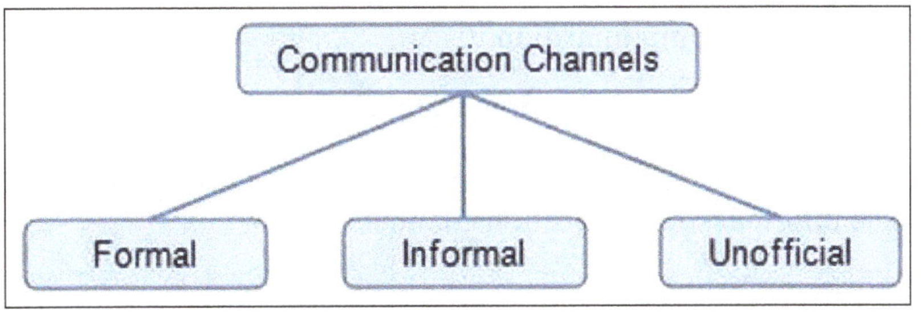

Importance of a Communication Channel

Communication channels affect how inefficient or efficient the flow of information is within a company. This lack of communication could cause employees to lack the knowledge of what the company expects of them, leaving them uninformed. Also without an effective communication channel, employees lose focus on the big picture and lose their company mind, which goes on to affects their decision making and productivity in the workplace.

This harms the overall organizational objectives as well. For an organization to be run effectively, a good manager should be able to communicate to their employees what is expected of them, make sure they are fully aware of company policies and any upcoming changes. Communication channels should be included by managers to enhance the productivity of their workers and to also ensure the well-being and smooth running of their company.

Types of Communication

With technology advancing, the number of communication channels has significantly increased over the past few years. Many new types of channels exist today including video conferencing, mobile technology, and electronic bulletin boards and also fax machines. As time progresses so many more channels will be introduced and implemented within the work place. These new types of channels help organizations communicate with the several workers to ensure their message has been sent. Choosing the type of communication channel is very important; these channels are grouped into three main categories: formal, informal and unofficial.

Formal Communication Channels:

- Sends information including goals, policies and procedures of an organization.
- Messages follow a chain of command.
- Flows from managers to their subordinates.
- Examples include business plan, customer satisfaction survey, annual reports, employer's manual, review meetings.

Informal Communication Channels:

- The strict hierarchical web of communication cannot function efficiently on its own and hence there exists a different communication channel.
- This type of communication channel may disrupt the chain of command, a good manager needs to find the fine balance between the formal and informal communication channel to ensure the best type of communication.
- Examples include, lunch time gathering, water cooler talk etc.
- Quality circles, team work, different training programs are outside of the chain of command are also informal communication channels.

Unofficial Communication Channels:

- Sometimes communication that takes place in an organization is interpersonal.
- Talks of sports, politics and TV shows are seen between co-workers.
- The unofficial communication channel in a company is the company's 'grapevine.' It is this that causes rumors.
- 'Grapevine' discussions often form groups, which translate into friendships outside of the organization.
- A good manager should be aware of information circulating in this unofficial communication channel and should take positive measures to prevent false information from spreading.
- Social gatherings between co-workers are a type of unofficial communication channel.

Transmission of Information

Communication theory studies methods for the design of messages for the transmission of information. As such, it is a branch of the subject of information theory. Communication theory is that part of information theory that is concerned with the explicit design of systems to convey messages and with the performance of those systems. Digital communication theory is that part of communication theory in which digital techniques play a prominent role in the communication process, either because the information to be transmitted is digital or because the information is temporarily represented in digital form for the purpose of transmission.

A message originating in an information source is to be transmitted to an information user through a channel. The digital communication system consists of a device called a transmitter, which prepares the source message for the communication channel, and a device called a receiver, which prepares the channel output for the user. The operation of the transmitter is called modulation or encoding. The operation of the receiver is called demodulation or decoding. At a fundamental level, a communication channel is normally an analog channel in that it transmits waveforms. Digital data must be modulated into an analog waveform for passage through the channel. The source data may arise as digital data or as analog data. The user may require the data be delivered in digital form or in analog form. When the source data is analog, perhaps a continuous- time analog waveform, a digital communication system converts it to digital data, processes it, and then converts it to a continuous-time analog waveform for passage through the channel. By means of sampling and quantization, the analog waveform out of the channel is converted to digital data for processing even though it may ultimately be required in analog form by the user. This constant conversion between digital data and analog data may seem inefficient but, because so much can be done to the data while it is in digital form, it is worthwhile. Besides, the analog waveform passing through the channel may be completely different from the analog waveform generated by the source.

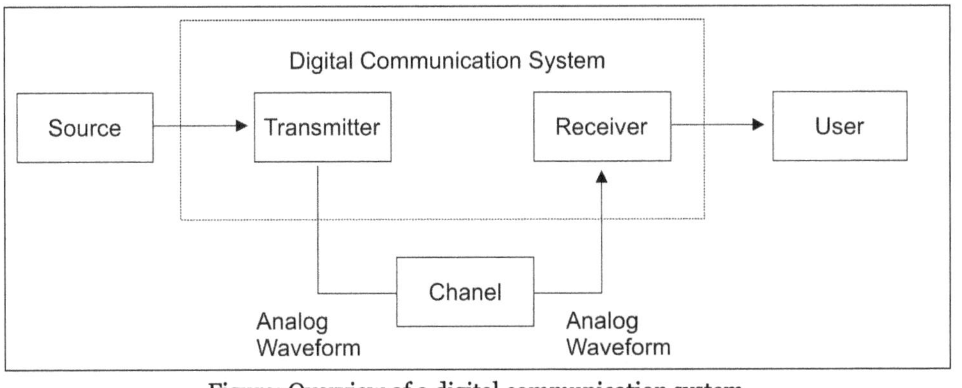

Figure: Overview of a digital communication system.

In contrast to digital modulation is analog modulation, which today is widely, used in radio, television, and phonography. Analog modulation techniques make relatively superficial changes to the signal in order to get it through the channel; there is no significant effort to tailor the waveform to suit the channel at any deeper level. Digital communication waveforms are much more sophisticated. Digital communication theory endeavors to find waveforms that are closely matched to the characteristics of the channel and that are tolerant of the impairments in the channel so that the reliable flow of information through the channel is ensured. The characteristics of the source are of no interest in designing good waveforms for a channel. Good waveforms for digital communication are designed to match the characteristics of the channel; the source information is then encoded into this channel waveform. A digital communication system might require a considerable amount of electronic circuitry to translate the source waveform into a form more suitable for the channel, but electronics is now cheap. In contrast, most channels are expensive and it is important to make the best use of a channel.

Even applications that might appear to be intrinsically analog applications, such as broadcast television, can be partitioned into two tasks — the task of delivering so many bits per second to the viewer and the task of representing the video signal by the available number of bits per second. This is the important and surprising separation principle of information theory, which says that the task of transmitting the output of a source through a channel can be separated, without loss, into the task of forming a binary representation of the source output and the task of sending a binary sequence through the channel. Otherwise there will be disadvantages such as increased bandwidth or larger transmitted power. The data must be compressed, and then modulated into a transmission waveform, perhaps one that carries multiple bits per second per hertz.

The only disadvantage of digital communication that must be accepted is that it is not always compatible with the existing analog communication receivers of earlier generations; it would be expensive and unpopular to discard existing AM and FM radio receivers in favor of a new high-fidelity digital format. Other disadvantages that are often cited are really not compelling; any validity that they once had has crumbled under the progress of technology. Specifically, cost was once a disadvantage of digital communication systems but no longer is. Bandwidth expansion is often mentioned as a disadvantage of a simple system, but with modern data compaction, and multilevel signaling, one can actually reduce the bandwidth by using digital communication.

Another disadvantage sometimes mentioned is that of quantization noise. Quantization noise, however, is completely under the control of the designer of the quantization scheme and will be the only important source of noise in the signal received by the user. The modern view is that quantization noise is a price cheerfully paid for the more important advantage of removing the effects of channel noise from the received signal. It is a truism of information theory that in a well-designed system the quantization noise will always be less than the channel noise it replaces.

On the other hand, there are numerous and compelling advantages of digital communication. Every link becomes simply a "bit pipe" characterized by its data rate and probability of bit error. It makes no difference to the link whether the bits represent digitized voice or computer programs. Many kinds of data source can share a common digital communication link, and the many kinds of communication link are suitable for every data source. A binary data stream can be routed through many physically different links in a complex system, and can be intermingled with other digital traffic in the network. A digital data stream is compatible with standardized encryption and anti-jam equipment. Errors due to noise and interference are almost completely suppressed by error correction codes.

The digital data stream can be regenerated at every repeater that it passes through so that the effect of additive noise does not accumulate in the signal. Analog repeaters, on the other hand, consist of amplifiers that amplify both signal and noise. Thus, noise accumulates as an analog signal passes through a series of repeaters. Finally, note that digital communication systems are built in large part from digital circuitry. A digital data format is compatible with the many other digital subsystems in a large system. Data can be readily buffered in random- access digital memories or on magnetic discs. Many functions of a modulator/ demodulator can be programmed into a microprocessor or designed into a special-purpose digital integrated circuit.

It is a curiosity of technological history that the earliest communication systems such as telegraphy were actually digital communication systems. The time-honored Morse code is a waveform for digital communication in that it uses a discrete alphabet. Telegraphy created a communications industry but lacked the popular appeal of later analog communication systems such as the telephone and the phonograph. The analog systems have been dominant for most of the twentieth century.

The earliest broadcast systems for communication were concerned with the transfer of analog continuous signals, first radio and then television signals. Analog modulation techniques were developed for embedding a continuous-time signal into a carrier waveform that could be propagated through a channel such as an electromagnetic wave channel. These techniques are still employed in systems that demand low cost or have strong historical roots, such as radio, telephony, and television, Indications can be seen, however, that analog modulation is becoming outdated even for those applications, and only the enormous investment in existing equipment will forestall the inevitable. Indeed the evolution from digital communication to analog communication that began in the 1870s is now being countered by an evolution from analog communication back to digital communication that found its strength in the 1970s. Even the analog phonograph record, after 100 years, has been superseded by the compact disk. The earliest radio transmitters used amplitude modulation. This form of analog modulation maps a signal s (t) into a waveform:

$$c\,(t) = \sin\left(2\pi f_0 + \int_0^t ms(\xi)\,d\xi\right)$$

Where m is a small constant called the modulation index such that ms (t) << 1, and where m is a constant called the carrier frequency such that fo is large compared to the largest frequency for which S(f), the Fourier transform of s(t), is nonzero. The mapping from s(t) to c(t) could be implemented in the transmitter very simply, and the inverse mapping from c(t) to s(t) could be implemented simply in the receiver, though only approximately. Therefore, amplitude modulation became very popular early on even though the fidelity of the received signal is not noteworthy.

Frequency modulation is an alternative analog modulation technique given by the following map from s (t) to c (t):

$$c\,(t) = \sin\left(2\pi f_0 + \int ms(\xi)\,d\xi\right)$$

Where, again, the carrier frequency fo is large compared to the largest frequency for which S(f) is significant. Frequency modulation was naively proposed very early as a method to conserve the radio spectrum. The argument was that the term ms(t) is an "instantaneous frequency" perturbing the carrier frequency fo and if the modulation index m is made very small, the bandwidth of the transform C(f) could be made much smaller than the bandwidth of S(f). Carson argued that this is an ill-considered plan, as is easily seen by looking at the approximation.

$$c(t) \approx \sin 2\pi f_0 t + \cos\, 2\pi f_0 t\left[m\int_0^t s(\xi)\,d\xi\right]$$

when m is small. The second term has the same Fourier transform as the bracketed component, but translated in frequency by f_0. Because the integral of s (t) has the same frequency components as s(t), the spectral width is not reduced. As a result of this observation, frequency modulation temporarily fell out of favor. Armstrong re-wakened interest in frequency modulation when he realized it had a much different property that was desirable. When the modulation index is large, the inverse mapping from the modulated waveform c(t) back to the signal s(t) is much less sensitive to additive noise in the received signal than is the case for amplitude modulation, at least when the noise is small. Frequency demodulation implemented with a hard limiter suppresses noise and weak interference and so frequency modulation has come to be preferred to amplitude modulation because of its higher fidelity.

The basic methods of amplitude modulation and frequency modulation are also used in modified forms such as single-sideband modulation or vestigial sideband modulation. These modified forms are attempts to improve the efficiency of the modulation waveform in its use of the spectrum. Other analog methods such as Dolby modulation are in use to match the analog source signal more closely to the noise characteristics of the channel. All of the methods for modifying the techniques of analog modulation are stopgap methods. They do not attack the deficiencies of analog modulation head on. Eventually, such methods will be abandoned for the most part in favor of digital modulation.

The theory of communication became much more mathematical when it became widely appreciated that communication is intrinsically a statistical process. There is no point in transmitting a predetermined message if the receiver already knows the message. If there are only a few possible predetermined messages already known to the receiver, one of which must be sent, then there is no need to send the entire message. Only a few prearranged bits need to be transmitted to identify the chosen message to the receiver, but this already implies that there is some uncertainty about which message will be the chosen message; randomness is an essential ingredient in communication because of the statistical nature of messages. Randomness is also an essential ingredient in communication because of noise in the channel. This statistical view of communication encompassing both random messages and noisy channels was popularized by Shannon. Earlier, Rice had made extensive study of the effect of channel noise on received analog communication waveforms. Shannon developed the broader and counterintuitive view that the waveform could be designed so as to make the channel noise essentially inconsequential in the quality of the received waveform. Shannon realized that combating noise was a job for both the transmitter and receiver, not for the receiver alone.

Shannon laid a firm foundation for the development of digital communication, A somewhat more applied paper, transitional between analog and digital communication, is due to Oliver, Pierce, and Shannon, This period seems to be the time when people were first thinking deeply about the fundamental problems and the return to digital signaling began, but there were many studies and applications of digital signaling earlier as in the work of Nyquist and Hartley. Aschoff gives a good history of digital signaling.

There is a misconception that is sometimes cited as a disadvantage of digital modulation; digital modulation of an analog signal is said to require larger bandwidth than direct analog modulation, this is not true. It was true when digital communication waveforms were limited to simple pulse code modulation. Today by using data compression, data compaction, and trellis-coded modulation that conveys multiple bits per second per hertz, one can have a digital communication system using less bandwidth than an analog communication system, and the fidelity of such systems is rapidly improving. The superiority of digital communication is more than just an empirical observation of recent engineering trends; its optimality is implicit in Shannon's original theorems.

Point-to-point Digital Communication

A simple block diagram of a point-to-point digital communication system was shown in figure. Even though we will use the terminology of communication theory to describe it, the model is quite general and can be applied to a variety of situations. One can interpret many other information-handling systems, such as mass storage systems, in terms of this model. It only is necessary to translate the terminology and to identify the properties of the channel. The identification of the channel may be somewhat arbitrary because some of the physical components such as amplifiers might sometimes be considered to be part of the channel and might sometimes be considered as part of the modulator and demodulator. The boxes labeled channel, data source, and data user are those parts of the system that are not under the control of the designer.

It is the task of the designer to connect the data source to the data sink by designing the boxes labeled transmitter and receiver. These boxes are also called, more simply, the encoder and decoder or the modulator and demodulator the latter names are usually preferred when the channel is a waveform channel while the former are usually preferred for discrete channels. Consequently, the transmitter is also called the encoder/modulator, and the receiver is also called the demodulator/decoder. Often a single device is designed so it can play either the role of the transmitter or the role of the receiver (or of both simultaneously); then the modulator and demodulator can be combined into a box called the modem. The term “modem” might also be construed to include the encoder and decoder as well, and might also include other functions that extend beyond modulation and demodulation. We prefer to use the terms “transmitter” and “receiver” as the broader terms that include supporting functions. Figures show the functions normally included in the transmitter and receiver.

The trend in the design of communication systems is to separate the design tasks associated with the data source and the data user from the design tasks associated with the channel. This leads technology in the direction of greater flexibility in that source data, when reduced to a stream of bits, might be transmitted through any one of several possible channels. To do this, the transmitter and receiver are broken into more detailed functions as described by the block diagrams of Figures. The transmitter includes a source encoder, a channel encoder, and a modulator; the receiver includes a demodulator, a channel decoder, and a source decoder, Information theory teaches us that there is no consequential loss in performance because of partitioning.the problem in this way. Moreover, there is no loss in generality if the interface between the source encoder and channel encoder and the interface between the channel decoder and source decoder are specified to be binary bit streams.

The source data may be analog or digital. Upon entering the transmitter, analog data must first be digitized. In the process of digitization, continuous time may be reduced to discrete time by the process of sampling of a source waveform of finite bandwidth. Then the data stream is processed by a source encoder, whose purpose is to represent the source data compactly by a stream of bits called the source codewords. At this point, the source data has been reduced to a commonplace stream of bits, superficially displaying no trace of the origin of the source data. Indeed, data from several completely different kinds of sources now may be merged into a single bit stream. The source data may then be encrypted to prevent eavesdropping by an unauthorized receiver. Again, the encrypted bit stream is another commonplace bit stream superficially displaying no trace of its origin.

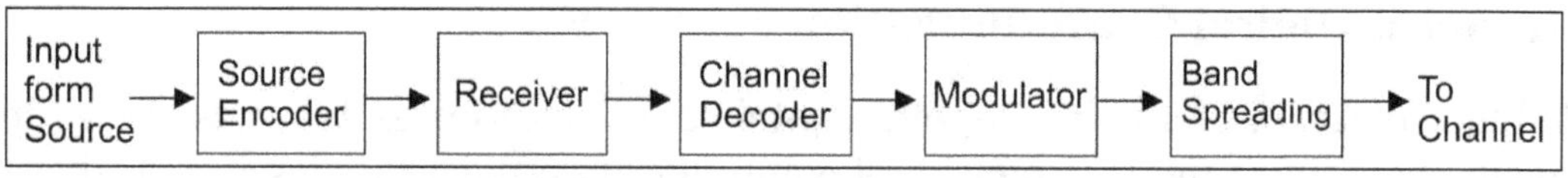

Figure: Primary transmitter functions.

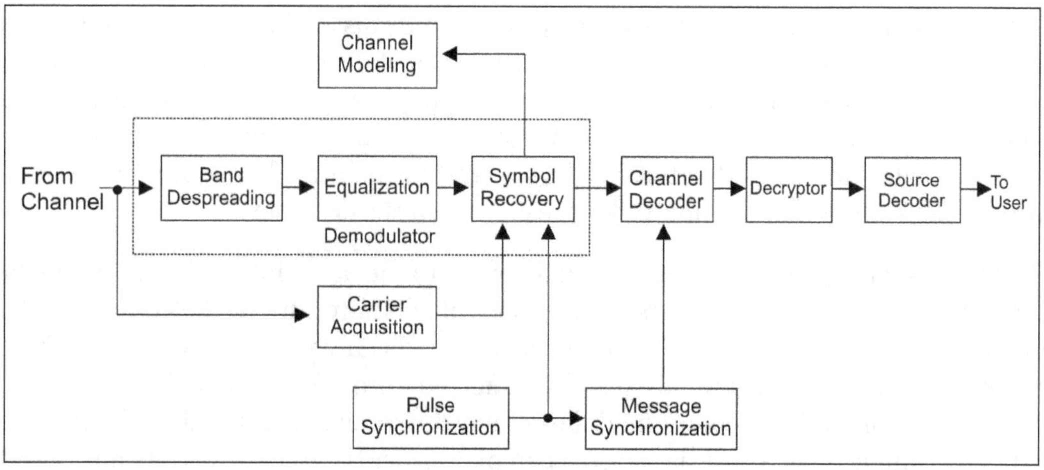

Figure: Primary receiver functions.

The bits are next processed by the channel encoder, which transforms the bit stream into a new data' stream called the channel codestream, the new data stream has more redundancy than the stream of source codewords; the purpose of the redundancy is to match the constraints of the channel and so that errors arising in the channel can be corrected. The symbols of the new data stream could be binary or could be symbols from a larger alphabet called the channel alphabet. The stream of channel codewords is passel to the modulator, which converts the sequence of discrete code symbols into a continuous function of time called the channel waveform. The modulator does this by replacing each symbol of the channel codeword by the corresponding analog symbol from a finite set of analog symbols composing the modulation alphabet. Here, discrete time is reconverted to continuous time. The sequence of analog symbols composes the channel waveform, which is transmitted through the channel directly or after the bandwidth is intentionally spread. The reason bandspreading might be used is to protect the signal from some kinds of interference, possibly intentional interference created by an adversary.

The input to the channel is the output of the transmitter. The channel waveform, as might be the source waveform, is now a continuous-time waveform. However, it will have an appearance and properties quite different from the source waveform that appeared at the input to the transmitter.

The input to the receiver is the output of the channel. Because the channel is subject to various types of noise, distortion, and interference, the waveform seen at the channel output differs from the waveform at the channel input. The waveform will always be subjected to thermal noise in the receiver, which is additive Gaussian noise, and this is the disturbance that we shall study most thoroughly. The waveform may also be subjected to many kinds of impulsive noise, burst noise, or other forms of non-Gaussian noise.

Upon entering the receiver, if the waveform has been band spread in the transmitter, it is first despread. The demodulator may then convert the received waveform into a stream of discrete channel symbols based oil a best estimate of each transmitted symbol. Sometimes it makes errors because the received waveform is not the same as was transmitted. Perhaps to quantify its confidence, the demodulator may append additional data to each demodulated symbol. The demodulated sequence of symbols, possibly including confidence annotations, is called the received word. The symbols of the received word do not always match those of the channel codeword. The function of the channel decoder is to use the redundancy in a channel codeword to correct the errors in the received word, and then to produce an estimate of the bit stream that appeared at the input to the channel encoder. If the bit stream has been encrypted, it is now decrypted to produce an estimate of the sequence of source codewords. Possibly at this point the bit stream contains source codewords from more than one source, and these source codewords must be deinterleaved. If all errors have been corrected by the channel decoder, each estimated source code word matches the original source codeword. The source decoder performs the inverse operation of the source encoder and delivers its output to the user.

The channel encoder and channel decoder are commonly split into two functions: that of implementing error control to negate the effects of channel noise; and that of preparing the sequence of transmitted symbols to be compatible with channel constraints. These functions are studied under the terms data transmission and data translation (or under the terms error-control coding and constrained channel coding). Data transmission encompasses the various discrete methods, such as the use of error-correcting codes, for combating noise and errors in the communication channel. Data translation encompasses various methods for preventing troublesome symbol patterns from appearing at the input to the modulator.

The modulator and the demodulator are studied under modulation theory. This is the core of digital communications. The main elements of the structure of the optimal receiver will be found with the aid of the maximum- likelihood principle.

The source encoder and source decoder are commonly split into two functions: that of expressing the output of the source compactly, and that of abridging the output of the source. We shall study these under the terms data compaction and data compression. Data compaction encompasses various methods for reducing redundancy in a stream of digital data. Data compression encompasses the various methods for reducing the entropy of a stream of data. This includes primarily methods such as quantization for digitizing analog signals. There are numerous parallels between codes for data compression and codes for data transmission. There are also numerous parallels between sampling theory, used to turn a continuous-time waveform into a discrete-time sequence, and modulation theory, used to turn a digital sequence into a continuous-time modulation waveform.

The receiver also includes other functions, such as equalization and synchronization shown in Figure that is needed to support demodulation. The structure of the optimal receiver will not fully emerge until these functions are developed later as a consequence of the maximum-likelihood principle. The receiver may also include intentional nonlinearities, perhaps meant to clip strong interfering signals, or to control dynamic range.

The functional anatomy of a digital communication system that we have sketched hem also serves as an outline for designing a communications system. The potency of the ideas that emerge may be underscored by mentioning how advantages may come froth unexpected directions. For example, the purpose of an error-control code may at first appear to be to improve the quality of the data, However, a subtler view is equally valid and coming into vogue. The error-control code can be thought of as a way to reduce transmitted power and antenna size in a modem digital communication system, this because the use of an error-control code allows the system to run at lower signal-to-noise ratio.

Networks for Digital Communication

In contrast to digital point-to-point communication systems are digital communication networks. Communication networks became important relatively recently, so they are mostly digital. The main exception to this is the telephone network. The telephone network is a kind of network in which switching (circuit switching) is used to create a temporary point-to-point communication channel (a virtual channel). A broadcast system, shown in figure, might also be classified as a communication network. However, if the same waveform is sent to all receivers, the design of a broadcast system is virtually the same as the design of a point-to-point communication system. Therefore, both the telephone network and a broadcast system can be seen as kinds of analog point-to-point communication systems. Most other communication networks are digital. Figure shows a modem form of a communication network, the multi- access communication network. It consists of a single channel, to which is attached a multitude of users who exchange messages. The rules by which the users take turns using the channel are called channel protocols. The protocol is the new element that makes a digital communication network more complicated than a digital point-to-point communication system.

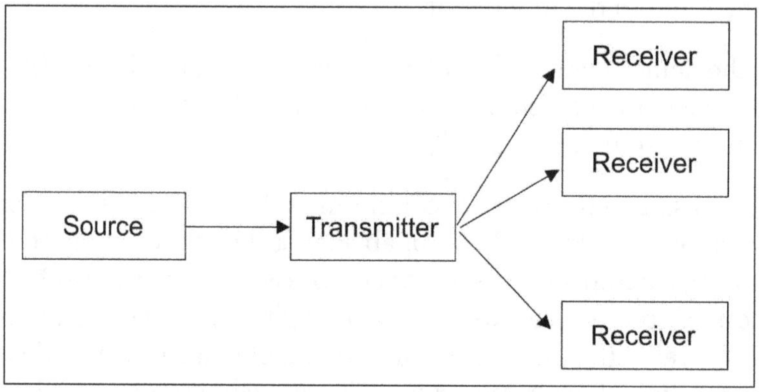

Figure: Broadcast communications.

The simplest multi-access channel is the fixed-assignment multi-access channel, of which lime-division multi-access signaling is a good example. Each user is pre-assigned a fixed time interval called a time slot during which it can transmit; one disadvantage of the time-division protocol is that it is inefficient; if a user has nothing to transmit; its time slot is wasted. Another disadvantage is that the set of users' needs to be fairly well specified so that time slots can be assigned. Each user, both transmitter and receiver, needs to know its slot assignment and needs to have its clock synchronized to system time.

At the other extreme are networks that take the form of a digital version of citizen's band radio. The set of users is very large, not very well defined, and each individual user transmits only rarely but at random times and at its own convenience. Then the channel is known as a demand-assignment multi-access channel. One kind of protocol for this case is the kind known as a contention resolution algorithm. Users transmit whenever it suits them but, in the event of a conflict, all users but one stop and retransmit at a later tithe. Algorithms to resolve contention are in wide use but are still incompletely understood. From a practical point of view one wishes to be ensured that a contention resolution algorithm will successfully deal with every possible situation of contention; otherwise the channel might fail because the protocol has deadlocked.

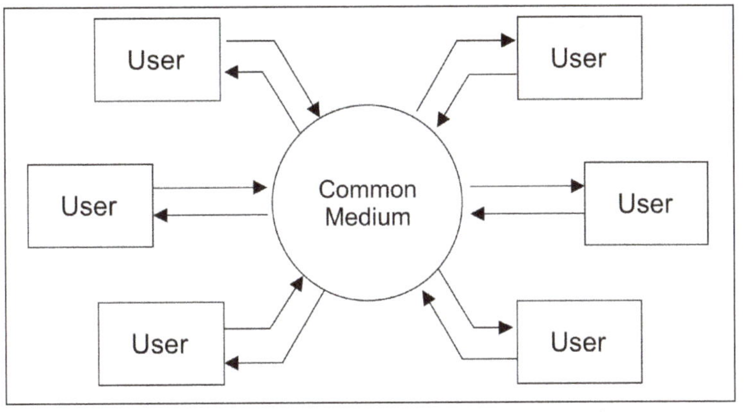

Figure: A multi-access communication network.

Multiplexed Channel

Channel multiplexing is the process of splitting or sharing the capacity of a high speed channel/ telecommunication link to form multiple low capacity/low speed sub-channels. A common kind of multiplexing merges a number of low-speed signals to send over an only high-speed link, or it is used to transmit a medium as well as its link with the number of devices. It provides both privacy & Efficiency. The entire process can be done using a device namely MUX or multiplexer, and the main function of this device is to unite n-input lines for generating a single output line. Thus MUX has many inputs & single output. A device is called DEMUX or Demultiplexer is used at the receiving end which divides the signal into its component signals. So it has single input and number of outputs.

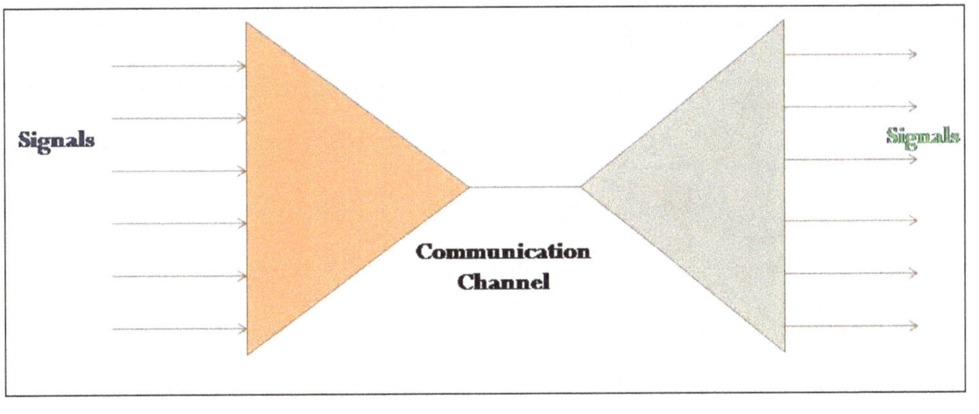

Types of Multiplexing Techniques

Multiplexing techniques are mainly used in communication, and these are classified into three types. The 3 types of multiplexing techniques include the following:

- Frequency Division Multiplexing (FDM).

- Wavelength Division Multiplexing (WDM).
- Time Division Multiplexing (TDM).

Frequency Division Multiplexing (FDM)

The FDM is used in telephone companies in the 20th century in long-distance connections for multiplexing number of voice signals using a system like a coaxial cable. For small distances, low-cost cables were utilized for different systems such as bell systems, K-and N-carrier; however, they don't let huge bandwidths. This is analog multiplexing used to unite analog signals. This type of multiplexing is useful when the link's bandwidth is better than the united bandwidth of the transmitted signals.

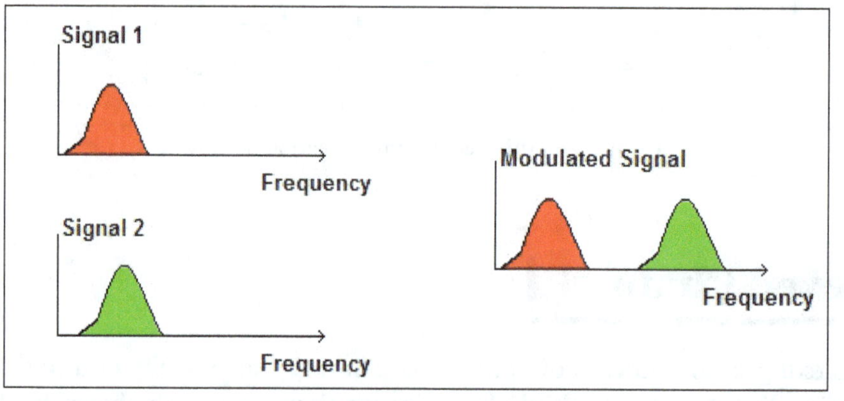

In FDM, signals are produced by transmitting various device modulated carrier frequencies, and then these are united into a solo signal which can be moved by the connection. To hold the adapted signal, the carrier frequencies are divided by sufficient bandwidth, & these ranges of bandwidths are the channels through the different traveling signals. These can be divided by bandwidth which is not used. The best examples of the FDM comprise signal transmission in TV and radio.

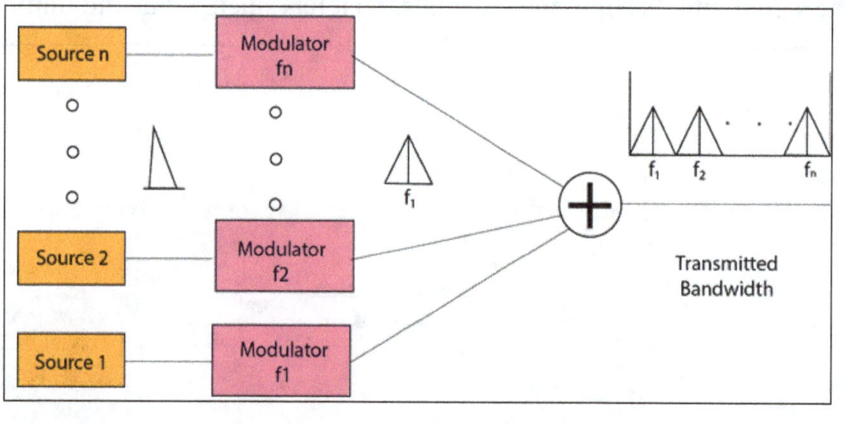

Advantages of FDM

- FDM is used for analog signals.
- FDM process is very simple and easy modulation.
- A Large number of signals can be sent through an FDM simultaneously.

- It does not require any synchronization between sender and receiver.

Disadvantages of FDM

- FDM technique is used only when low-speed channels are required.
- It suffers the problem of crosstalk.
- A Large number of modulators are required.
- It requires a high bandwidth channel.

Applications of FDM

- FDM is commonly used in TV networks.
- It is used in FM and AM broadcasting. Each FM radio station has different frequencies, and they are multiplexed to form a composite signal. The multiplexed signal is transmitted in the air.

Wavelength Division Multiplexing (WDM)

In fiber communications, the WDM (Wavelength Division Multiplexing) is one type of technology. This is the most useful concept in high-capacity communication systems. At the end of the transmitter section, the multiplexer is used to combine the signals as well as at the end of receiver section, de-multiplexer for dividing the signals separately. The main function of WDM at the multiplexer is for uniting various light sources into an only light source, and this light can be changed into numerous light sources at the de-multiplexer.

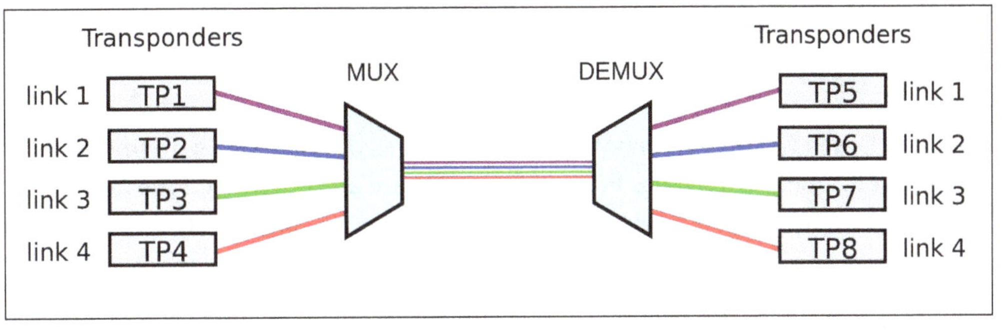

The main intention of WDM is to utilize the high data rate capacity of the FOC (fiber optic cable). The high data rate of this FOC cable is superior to the data rate of the metallic transmission cable. Theoretically, the WDM is similar to the FDM, apart from the data transmission through the FOC in which the multiplexing & de-multiplexing occupies optical signals.

Time Division Multiplexing (TDM)

The Time division multiplexing (or) TDM is one kind of method for transmitting a signal over a channel of particular communication with separating the time edge into slots. Like single slot is used for each message signal.

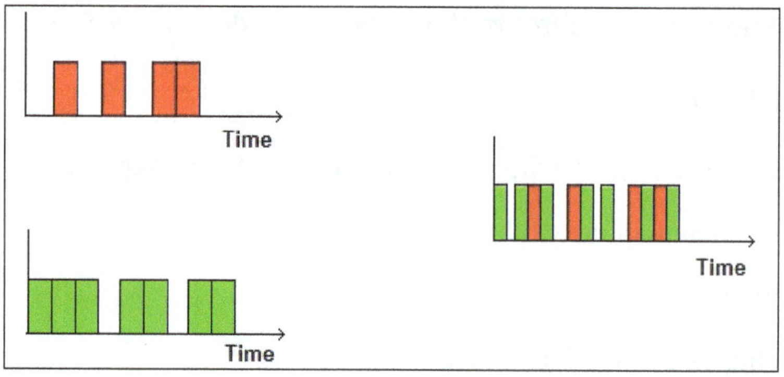

TDM is mainly useful for analog and digital signals, in which several channels with low speed are multiplexed into high-speed channels used for transmission. Depending on the time, every low-speed channel will be assigned to an exact position, wherever it works in the mode of synchronized. Both the ends of MUX and DEMUX are synchronized timely & at the same time switch toward the next channel.

Types of Time Division Multiplexing

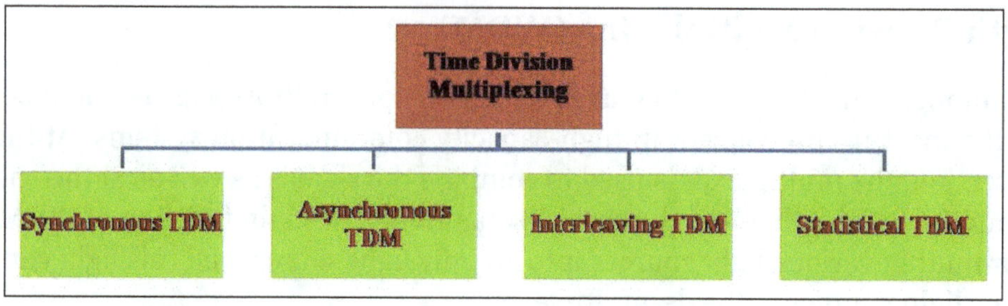

The different types of TDM include the following:

Synchronous TDM

The synchronous TDM is very useful in both analog as well as digital signals. In this type of TDM, the connection of input is allied to a frame. For example, if there are n-connections in the frame, then a frame will be separated into n-time slots, and for every unit, each slot is assigned to every input line.

In the sampling of synchronous TDM, the speed is similar for every signal, as well as this sampling needs a clock (CLK) signal at both the ends of sender & receiver. In this type of TDM, the multiplexer assigns the similar slot for each device at every time.

Concept of Synchronous TDM

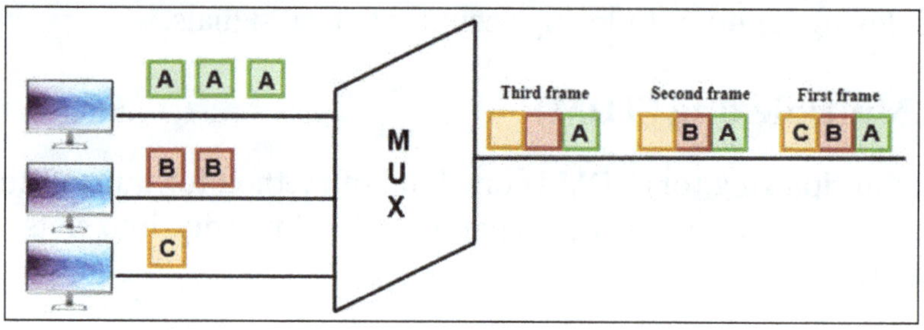

In the above figure, the Synchronous TDM technique is implemented. Each device is allocated with some time slot. The time slots are transmitted irrespective of whether the sender has data to send or not.

Disadvantages of Synchronous TDM

- The capacity of the channel is not fully utilized as the empty slots are also transmitted which is having no data. In the above figure, the first frame is completely filled, but in the last two frames, some slots are empty. Therefore, we can say that the capacity of the channel is not utilized efficiently.
- The speed of the transmission medium should be greater than the total speed of the input lines. An alternative approach to the Synchronous TDM is Asynchronous Time Division Multiplexing.

Asynchronous TDM

In asynchronous TDM, for different signals, the rate of sampling is also different, and it doesn't need a general clock (CLK). If the device has nothing for transmitting, then the time slot is assigned to a new device. The design of a commutator otherwise de-commutator is not easy & the bandwidth is low for this type of multiplexing, and it is applicable for not synchronous transmit form network.

Concept of Asynchronous TDM

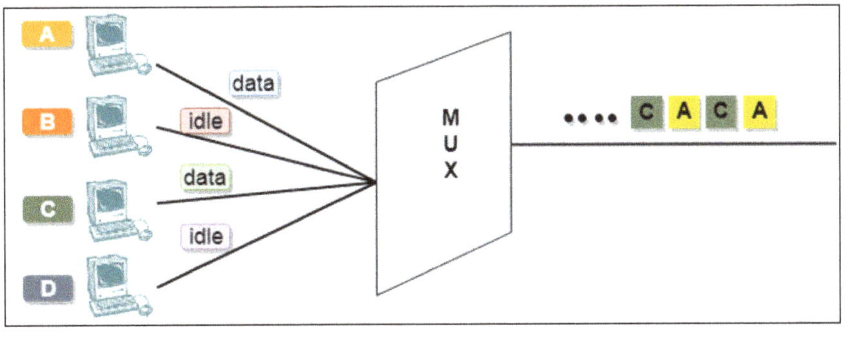

In the diagram, there are 4 devices, but only two devices are sending the data, i.e., A and C. Therefore, the data of A and C are only transmitted through the transmission line.

Frame of previous diagram can be represented as:

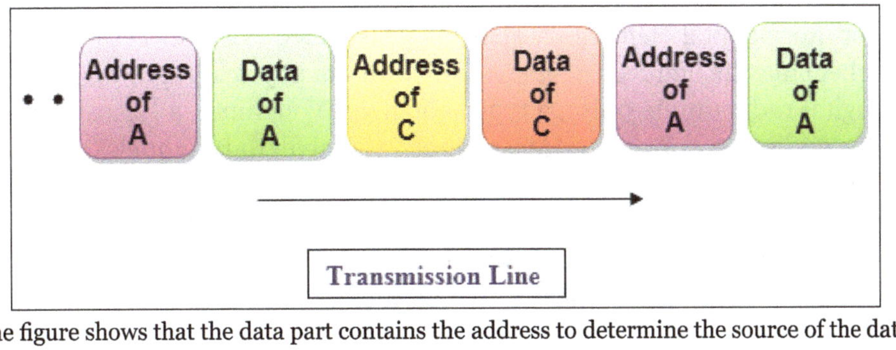

The figure shows that the data part contains the address to determine the source of the data.

Interleaving TDM

The TDM can be imagined like two speedy rotary switches on the multiplexing & demultiplexing surface. These switches can be rotated & synchronized in reverse directions. Once the switch releases at the surface of multiplexer ahead of a connection, then it has a chance of sending a unit into the lane. Similarly, once the switch releases at the surface of de-multiplexer ahead of a connection a chance to receiving a unit from the lane. This procedure is named as interleaving.

Statistical TDM

The statistical TDM is applicable to transmit different types of data simultaneously across a single cable. This is frequently used to handle data being transmitted through the network like LAN (or) WAN. The transmission of data can be done from the input devices which are connected to networks like computers, fax machines, printers, etc. The statistical TDM can be used in the settings of telephone switchboards to control the calls. This type of multiplexing is comparable to dynamic bandwidth distribution, and a communication channel is separated into a random data stream number.

Applications of Multiplexing

The applications of multiplexing include the following:

- Analog Broadcasting
- Digital Broadcasting
- Telephony
- Video Processing
- Telegraphy

Virtual Channel

A virtual channel (VC) is a communications track between two nodes giving the bandwidth needed for a virtual connection across the network. A virtual channel is the designation, usually a number that is recognized by the user as the single entity that will provide access to an analog TV program or a set of one or more digital elementary streams. It is called "virtual" because its identification (name and number) may be defined independently from its physical location. Examples of virtual channels are: digital radio (audio only), a typical analog TV channel, a typical digital TV channel (composed of one audio and one video stream), multi-visual digital channels (composed of several video streams and one or more audio tracks), or a data broadcast channel (composed of one or more data streams). In the case of an analog TV channel, the virtual channel designation will link to a specific physical transmission channel. In the case of a digital TV channel, the virtual channel designation will link both to the physical transmission channel and to the particular video and audio streams within that physical transmission channel.

Virtual Channel Flow Control

A virtual channel consists of a buffer that can hold one or more flits of a packet and associated state information. Several virtual channels may share the bandwidth of a single physical channel.

Virtual channels decouple allocation of buffers from al location of channels by providing multiple buffers for each channel in the network. If a blocked packet A holds a buffer *bio* associated with channel *Ci,* another buffer bi 1 is available allowing other packets to pass A. Figure below illustrates the addition of virtual channels to the network of Figure below. Packet A remains blocked holding buffers 3E.1 and 4S.l. In Figure below, however, Packet B is able to make progress because buffer 3E.2 is available allowing it access to channel (3E to 4W).

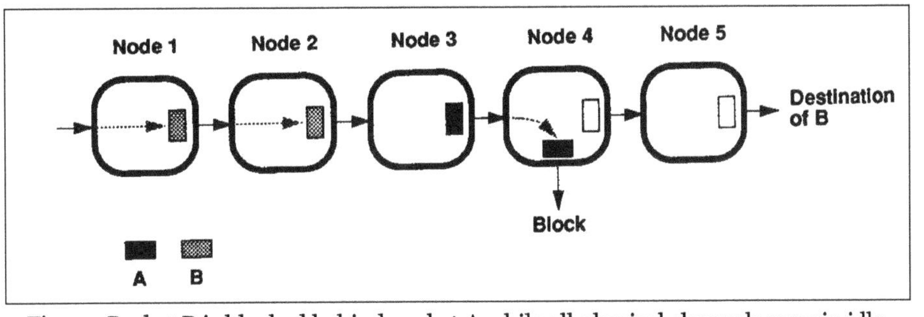

Figure: Packet B is blocked behind packet A while all physical channels remain idle.

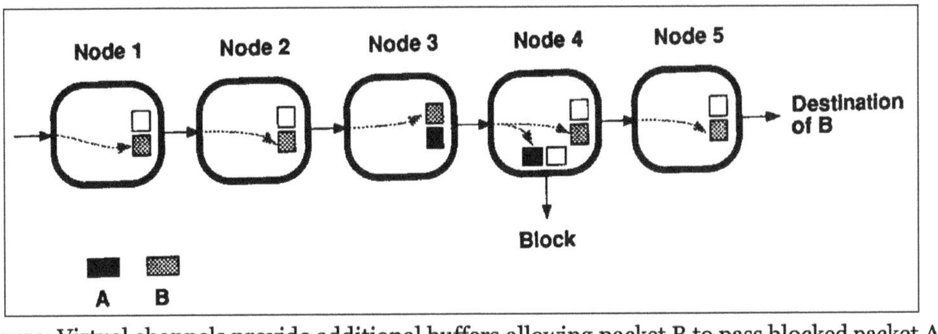

Figure: Virtual channels provide additional buffers allowing packet B to pass blocked packet A.

Adding virtual channels to an interconnection network is analogous to adding lanes to a street network. A network without virtual channels is composed of one-lane streets. In such a network, a single blocked packet blocks all following packets. Adding virtual channels to the network adds lanes to the streets allowing blocked packets to be passed.

Virtual channels provide an additional degree of freedom in allocating resources to packets in the network. This flexibility permits the use of scheduling strategies, such as routing the oldest packet first, that reduce the variance of network latency.

The most costly resource in an interconnection network is physical channel (wire) bandwidth. The second most costly resource is buffer memory. Adding virtual channel flow control to a network makes more effective use of both of these resources by decoupling their allocation. The only expense is a small amount of additional control logic.

The use of virtual channels for flow control builds on work in using virtual channels for deadlock avoidance and in using output queuing or split input queues for partial resource decoupling. A cyclic network can be made deadlock-free by restricting routing so there are no cycles in the channel dependency graph and then adding virtual channels to reconnect the network. Virtual channels were first implemented for this purpose in the torus routing chip.

The network design frame and the J-Machine network use virtual channels to provide two logical networks on a single physical network. The iWARP processing element uses virtual channels (called logical channels) primarily to guarantee bandwidth to virtual circuits. iWARP virtual channels are sufficiently general that they can be used to decouple resource allocation as described here.

A single stage of resource decoupling is provided by output queuing. By performing the queuing in the output of a switch rather than the input, arriving packets are able to pass blocked messages arriving on the same input. Tamir has shown how to achieve the same single-stage resource decoupling by partitioning the switch's input queue. This single stage resource decoupling is effective only if an entire packet fits in a single node. When a packet too long to fit entirely in one input queue is blocked, it backs up into the output stage of the previous node preventing any following packet from passing it. With output queuing, there is still only a single output buffer associated with each physical channel. If a packet blocks while holding this output buffer, the channel is idled.

Preliminaries

Topology

An interconnection network consists of a set of nodes, N and a set of channels, $C \subseteq N \times N$. Each channel is unidirectional and carries data from a source node to a destination node. A bidirectional network is one where (nl, n2) $\in$ C => (n2, n l) $\in$ *C*.

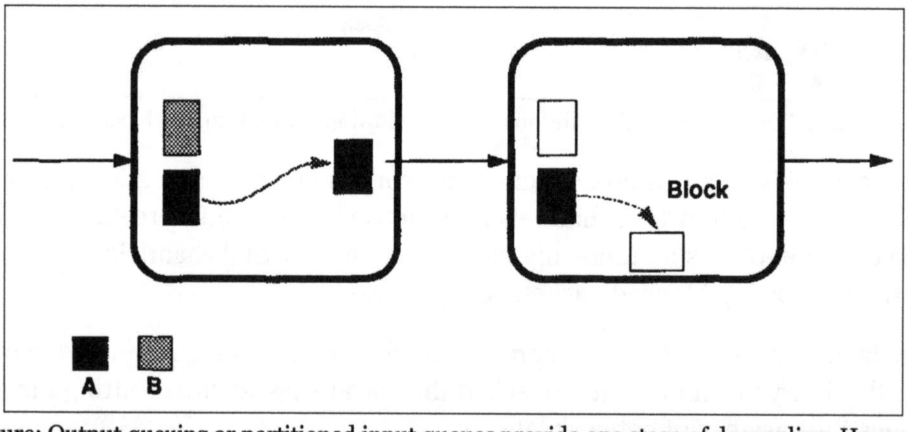

Figure: Output queuing or partitioned input queues provide one stage of decoupling. However, long packets (such as packet A) continue to couple resources and cannot be passed.

We have analyzed the performance of virtual channel flow control on multistage, k-ary n-fly, networks and have simulated the use of virtual channels on multistage, networks and direct, k-ary n-cube, networks.

Multistage (k-ary n-fly) networks have k^n inputs connected to *kn* outputs by n-stages of kn^{-1} $k \times$ k-switches. For example, a 2-ary 4-fly is shown in figure below.

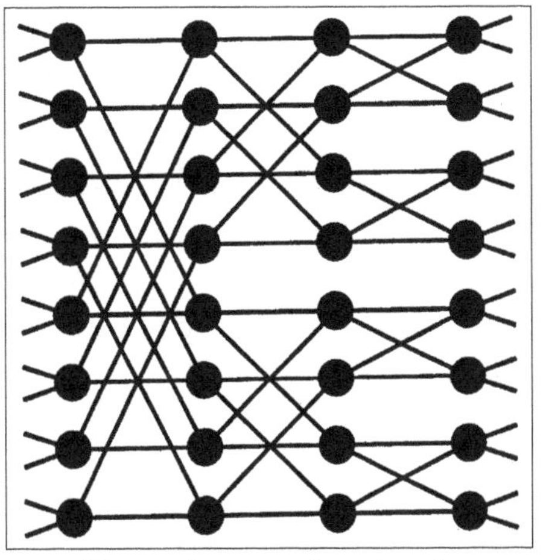

Figure: A 2-ary 4-fty network.

A k-ary n-cube mesh network consists of k^n nodes arranged in an n-dimensional grid. Each node is connected to its Cartesian neighbors in the grid. For example, a 16-ary 2-cube is shown in figure below.

Figure: A 16-ary 2-cube network.

The use of virtual channel flow control is in no way restricted to these two classes of networks. It is equally applicable to other topologies including trees, sorting networks, and irregular structures.

Routing

A packet is assigned a route through the network according to a *routing relation,* $R \subseteq C \times N \times C$, given the channel occupied by the head of the packet and the destination node of the packet, the routing relation specifies a (possibly singleton) set of channels on which the packet can be routed.

Flow Control

Communication between nodes is performed by sending *messages.* A message may be broken into one or more *packets* for transmission. A packet is the smallest unit of information that contains routing and sequencing information. A packet contains one or more flow control digits or *flits*. A flit is the smallest unit on which flow control is performed. Information is transferred over physical channels in physical transfer units or *phits*. A phit is usually the same size or smaller than a flit.

The flow control protocol of a network determines 1) how resources (buffers and channel bandwidth) are allocated and 2) how packet collisions over resources are resolved. A resource collision occurs when a packet P is unable to proceed because some resource it needs (usually a buffer) is held by another packet. Collisions may be resolved by 1) blocking P in place, 2) buffering P in a node prior to where the collision occurs, 3) dropping P, or 4) misrouting P to a channel other than the one it requires.

The flow control strategy allocates buffers and channel bandwidth to flits. Because flits have no routing or sequencing information, the allocation must be done in a manner that keeps the flits associated with a particular packet together. This may be done by associating a set of buffers and some control state together into a virtual channel. A virtual channel is allocated in a packet and the buffers of the virtual channel are allocated in a FIFO manner to the flits of that packet. Most networks associate only a single virtual channel with each physical channel.

Wormhole Routing

The technique described here is particularly suitable for use in networks that use wormhole routing. Wormhole routing refers to a flow-control protocol that advances each flit of a packet as soon as it arrives at a node (pipelining) and blocks packets in place when required resources are unavailable. Wormhole routing is attractive in that 1) it reduces the latency of message delivery compared to store and forward routing, and 2) it requires only a few flit buffers per node. Wormhole routing differs from virtual cut-through routing in that with wormhole routing it is not necessary for a node to allocate an entire packet buffer before accepting each packet. This distinction reduces the amount of buffering required on each node making it possible to build fast, inexpensive routers.

Virtual Channel Flow Control

Structure

Each node of an interconnection network contains a set of buffers and a switch. We assume that the buffers are partitioned into sets associated with each input channel, an input-buffered node, as shown in figure. An output-buffered switch can be considered to be an input buffered switch with a non-blocking first stage by associating the buffers on the output of each stage with the inputs of the next stage.

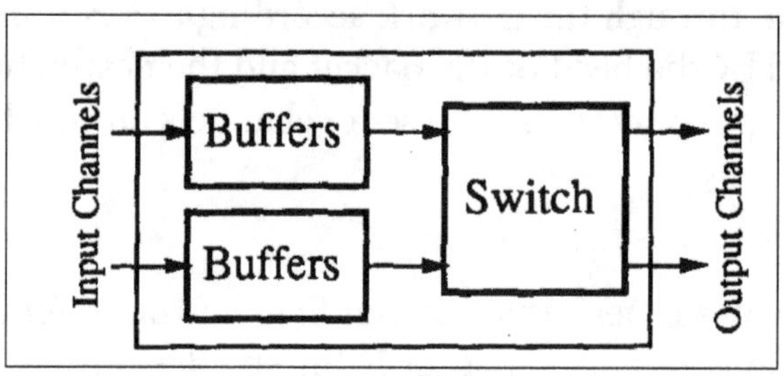

Figure: Node organization. Each network node contains a set of buffers for each input channel and a switch.

A conventional network organizes the flit buffers associated with each channel into a first-in, first-out (FIFO) queue as shown in figure. This organization restricts allocation so that each flit buffer can contain only flits from a single packet. If this packet becomes blocked, the physical channel is idled because no other packet is able to acquire the buffer resources needed to access the channel.

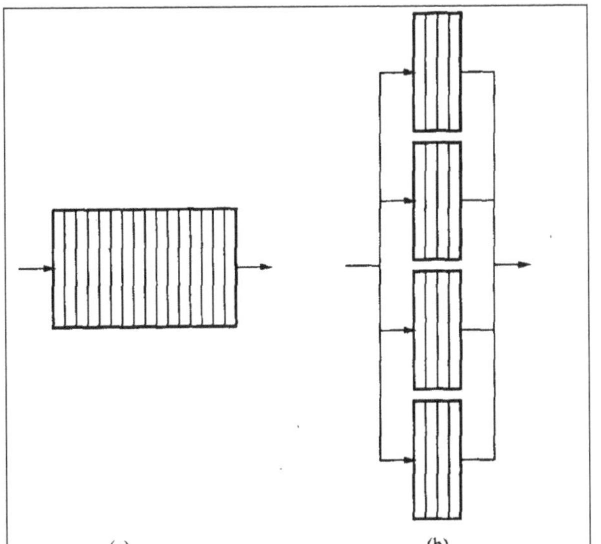

Figure: (a) Conventional nodes organize their buffers into FIFO queues restricting routing.
(b) A network using virtual-channel flow control organizes its buffers into several independent lanes.

A network using virtual channel flow control organizes the flit buffers associated with each channel into several lanes. The buffers in each lane can be allocated independently of the buffers in any other lane. This added allocation flexibility increases channel utilization and thus throughput. A blocked message, even one that extends through several nodes, holds only a single lane idle and can be passed using any of the remaining lanes.

Operation

In a network using virtual channel flow control, flow control is performed at two levels. Virtual channel assignment is made at the packet level while physical channel bandwidth is allocated at the flit level. When a packet arrives at a node, it is as signed (according to the routing algorithm) to an output virtual channel. This assignment remains fixed for the duration of the packet. The virtual channels associated with a physical channel arbitrate for physical channel bandwidth on a flit-by-flit basis.

Figure illustrates the hardware required to support virtual channel flow control on one physical channel. The transmitting node (node A) contains a status register for each virtual channel that contains the state of the lane buffer on the receiving node (node B). This state information includes: a bit to indicate if the lane is free, a count of the number of free flit buffers in the lane, and optionally the priority of the packet occupying the lane. Node B contains a lane buffer and a status register for each virtual channel. The status maintained on node B includes input and output pointers for each lane buffer and the state of the channel: free, waiting (to be assigned an Output), and active.

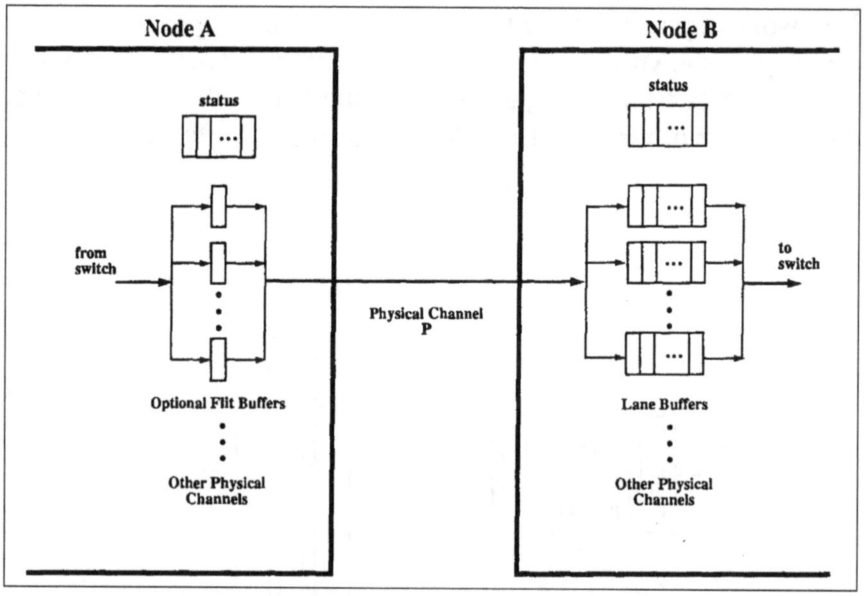

Figure: Logic associated with one physical channel P to support virtual channel flow control. The transmitting node (node A) includes status registers for each virtual channel and optional flit buffers. The receiving node (node B) contains buffers and status registers.

Lane assignment for physical channel P is performed by node A. When a packet arrives in an input buffer on node A not shown), it is assigned a particular output channel based on its destination, the status of the output channels, and the routing algorithm in use. The flow-control logic then assigns this packet to any free lane of the selected channel. If all lanes are in use, the packet is blocked in the waiting state until a lane is available. Maintaining lane state information on the transmitting end of the channel allows lane assignment to be performed on a single node. No additional internode communication is required to maintain this information as it is already required for flit-level flow control.

Once a lane is assigned to a packet, flit-level flow control is used to advance the packet across the switch and physical channel. To advance from an input buffer on the node A to an input buffer on node B, a flit must gain access to 1) a path through the switch to reach the output of node A, and 2) the physical channel to reach the input of node B. Typically either the switch is non-blocking, and thus always available, or a few optional flit buffers are provided at the output of node A so that switch and channel resources do not have to be allocated simultaneously.

When the last flit of a message (the tail flit) leaves a node the lane assigned to that packet is de-allocated and may be reassigned to another packet. The status register storage required to implement virtual channel flow control for one physical channel, S_{vc} is shown below in terms of the number of lanes, l, the total number of flit buffers in the receiver, b, and the number of bits used to encode priority, pri. Setting l = 1 and pri = O gives the status register storage required by a conventional channel S_{conv}. The first term of the expression corresponds to the storgecn. node A while the second term describes the storage on node B. For typical values of b = 16, l = 4, and pri= 0, S_{vc} is 36 bits compared with 17 bits for S_{conv}. This overhead is small compared to the storage required for the lane buffers, 512 bits 1f the flit size is 32 bits.

$$S_{vc} = l(\lg\left(\frac{b}{l}\right) + 1 + \text{pri}) + l\left(2\lg\left(\frac{b}{l}\right) + 2\right)$$

Allocation Policies

Flit-level flow control across the physical channel involves allocating channel bandwidth among lanes that 1) have a flit ready to transmit and 2) have space for this flit at the receiving end. Any arbitration algorithm can be used to allocate this resource including random, round-robin, or priority. For each physical channel, the arbitration algorithm is implemented as combinational logic that operates on the contents of the status registers and picks the highest priority lane that has space available at the receiving end. For random and round-robin arbitration schemes, priority information is generated by logic based on the lane's position and the previous state. For priority based schemes, priority information is stored in the status register for each lane.

Deadline scheduling can be implemented by allocating channel bandwidth based on a packet's deadline or age-earliest deadline or oldest age first. Scheduling packets by age reduces the variance of message latency. Deadline scheduling provides several classes of delivery service and reduces the variance within each class.

Implementation Issues

Virtual channel flow control can be integrated into existing switch designs by replacing FIFO buffers with multilane buffers. When this replacement is made, however, the switch must be modified to deal with a larger number of inputs and outputs, and the flow control protocol between nodes must be modified to identify lanes.

Increasing the number of virtual channels multiplexed on each physical channel increases the number of inputs and outputs that must be switched at each node. If the switch handles each of these inputs and outputs separately, the switch complexity will increase significantly. Increasing the switch complexity is not required, however. The average data rate out of the set of lanes associated with a given physical channel is limited to the bandwidth of the channel. Thus it is sufficient to provide a single switch input for each physical input and output channel. With this organization, a small (one or two flit) output buffer is desirable to decouple switch allocation from physical channel allocation. Individual lanes are multiplexed onto the single path through the switch in the same manner that they are multiplexed over the single physical channel between the nodes.

An intermediate organization is to provide a switch with separate inputs and multiplexed outputs. This configuration has the advantage of simple arbitration. An input virtual channel competes only for switch output ports. It need not simultaneously arbitrate for both input and output ports as is required for a fully multiplexed switch.

Any network that uses blocking or buffering flow control must, for each channel, send information in the reverse direction to indicate the availability of buffering on the receiving node. These acknowledgment signals can be transmitted on separate wires or, in a bidirectional network; they can be transmitted out of band on a channel in the opposite direction.

In a network using multilane buffers, two effects increase the acknowledgment traffic. First, a few bits must be added to each acknowledgment signal to identify the lane being acknowledged. Second, because a lane buffer is typically smaller than a channel FIFO, the use of block acknowledgments to amortize the cost of the signal over several flits is restricted.

Even with these effects, acknowledgment signal bandwidth is still a small fraction of forward bandwidth. In a network with 32-bit flits, 15 lanes per channel, and no block acknowledgment, 4 bits must be sent along the reverse channel for each flit transmitted along the forward channel, a 12.5% overhead. An additional 12.5% overhead is required to identify the lane associated with each flit sent in the forward direction. Such a scheme could be realized by a physical channel consisting of a 9-bit forward path (8-bit phits) and a 1-bit reverse path. Every four channel cycles a 32-bit flit is transmitted over the forward path along with its 4-bit lane identifier and a 4-bit acknowledgment code is transmitted over the reverse path.

Return Channel

In a duplex transmission channel, return channel is sometimes the case that the main channel operates only in one direction (i.e. simplex), but that a channel of much lower capacity (and much lower cost) operates in the opposite direction: this is the return channel. It is chiefly used for monitoring the main channel, and for notifying the transmitter of errors detected by the receiver on the main channel.

Audio Return Channel (ARC)

ARC or Audio Return Channel is a special and dedicated channel for passing the audio signals back to the sound device over HDMI. Simply, we can say if a television supports ARC, it can pass (upstream/ downstream) or return the TV's sound to the connected audio device over the HDMI cable. Nowadays ARCs gaining its popularity and as a result, most of the entertainment devices such as TVs, HT systems, Sound bars are coming with this feature.

How HDMI Audio Return Channel Works?

Audio Return Channel is a cool HDMI feature and no doubt it is beneficial. Suppose you have three devices TV, DTH & a Sound Bar. And the scenario is DTH is connected to with the TV and the TV is connected with the sound bar with HDMI ARC port.

So, in this case, as sound bar is not directly connected with the DTH, but can even play the sound of DTH content with the help of HDMI Audio Return Channel. Isn't interesting? So, how is that works? Having ARC and HDMI CEC (optional) are enabled, your TV will automatically detect if any active audio system either connected with the ARC port or not. If yes, the television starts passing (down streaming) the audio to that connected sound system instead of playing this by itself. If you turn off the sound system, the TV then automatically starts playing the sound by itself (if HDMI CEC is enabled in TV) and stops passing (down streaming) the audio to the sound system.

Setting up

Well, it's pretty easy but not a single step. However, no of steps depends on your TV and audio system. Please note, to use ARC features, you must have ARC enabled TV, as well as ARC, enabled Sound system.

Step 1: Connection over ARC

Connect TV and Sound system with an HDMI. You need to remember the cable should be connected into ARC enabled port on both devices.

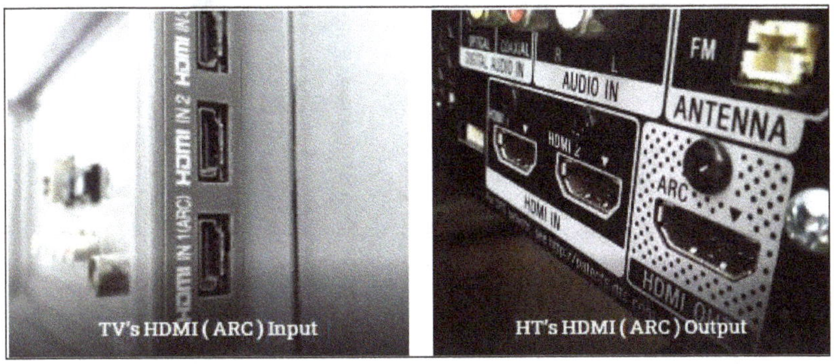

Step 2: Connection over Non-ARC

Now, connect remaining input devices like DTH box or Blu-ray Player into vacant HDMI ports either on the sound system or TV. In case you connect the remaining devices to the sound system, then you need to turn on the sound system. But connecting all devices directly into the TV, you will get the actual flavor of ARC. That means, while your sound system is on, you will receive the audio from the sound system otherwise TV will play the sound by itself.

Step 3: Choose Audio Out

Go to the TV's audio settings and select 'External Speaker over the ARC/ Optical' option from 'sound output'.

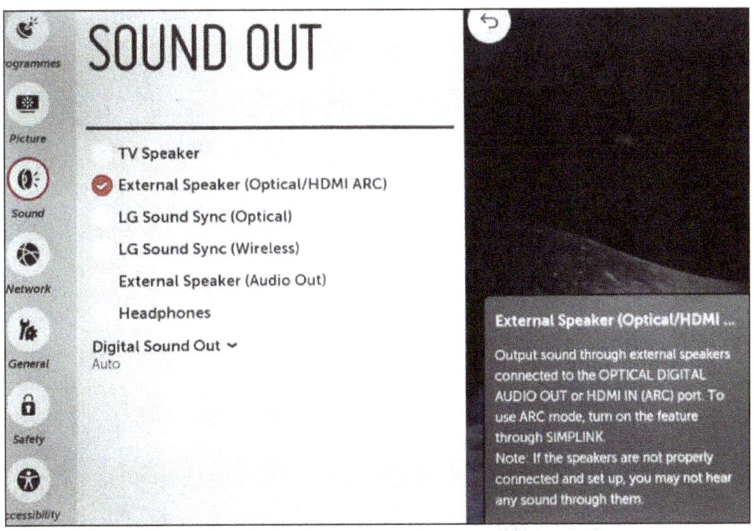

Step 4: Enable HDMI CEC

Enable HDMI CEC from TV settings. However, this option is not mandatory but can be helpful to get the optimum result of ARC.

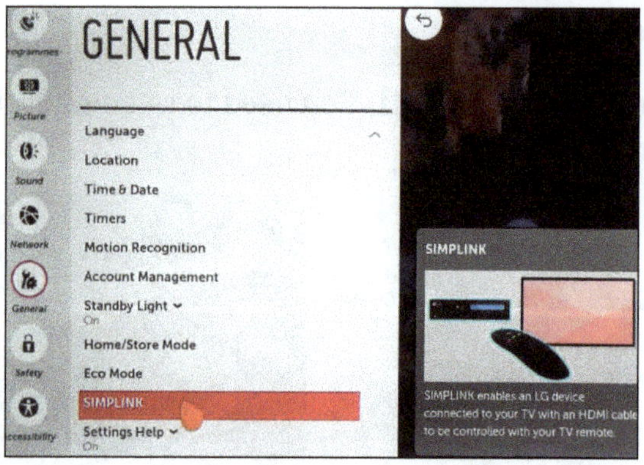

Applications and Benefits of ARC

(Non-arc inter-connection) shows all A/V sources are directly connected with an A/V receiver or a giant sound system. This means you cannot play any devices without the A/V receiver. You may say, so what's the problem! The problem is, budget home theaters or sound bars do not have enough HDMI ports to connect all A/V devices. So, this means, you cannot enjoy audio from the sound system with other devices.

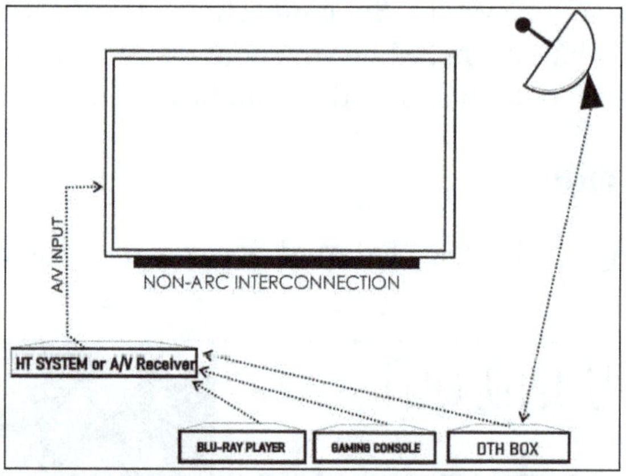

Unavailability of HT will stop playing audio.

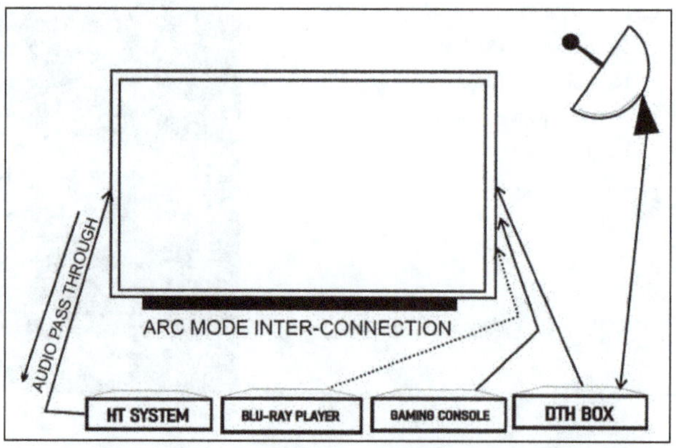

Unavailability of HT will make audio from TV's speaker.

On the other hand, in figure above (Arc mode inter-connection), the tv is connected with all devices including the Home Theatre or A/V receiver. The TV and the A/V receiver is connected exclusively over the ARC HDMI port. Now the miracle starts. With the help of ARC, the sound from any of the connected sources will be played by the Home Theatre system. Now if you turn off the HT system, then the sound will be played by the TV. So, now you can use just a sound bar to play the audio from various inputs.

Disadvantages

ARC is a technology revolution that simplifies home entertainment, although it has a few drawbacks. However, it's not the deal-breaker but it's good to know about.

ARC can Pass up to 5.1 Channel Audio

ARC can return or pass both uncompressed and compressed audio data to the sound system. PCM (2 channel) & LPCM (multi-channel) are the most common types of uncompressed audio formats. Dolby Digital & DTS is an audio encoder for compressing the audio that uses multiple audio channels which provides a surround sound. You should remember that most TVs bundled with ARC can pass only 2.1 (PCM) stereo audio data, whatever your inputs are. Only some premium TVs can pass up to 5.1 Dolby Digital or DTS-ES audio to the sound system. ARC is not capable to return Dolby Digital Plus, Dolby Atmos and DTS: X encoded audios. Dolby Atmos and DTS: X are premium audio surrounding technologies and those are mostly available on 7.1 channel. However, eARC (Enhanced Audio Return Channel) solves these issues but as a new technology, it's coming with premium televisions. There are some misconceptions about HDMI audio return channel cable. Please note no special HDMI cable is required to get this feature. Only eARC (Enhanced Audio Return Channel) required HDMI 2.1 cable.

The Bottom Line

ARC is a great feature and comes with most of modern TVs and Sound Systems. Now a days without ARC based TV is quite impossible. But make sure that the TV supports 5.1 channel ARC which will be the best if you have a 5.1 Home Theatre System. Another important thing, always try to purchase TV having the HDMI CEC feature to get best experience.

Upstream Channel and Downstream Channel

Upstream

Upstream or upload speed is the term given to the data that your computer sends out to the Internet. Your computer uploads data anytime you send an email or post a picture to Facebook or a server using FTP. Generally, the upstream speed capable via cable modems ranges from 400 kilobits per second to 600 kilobits per second (Kbps). This represents the total bandwidth of data your computer can send out to the Internet each second at full speed.

Downstream

Downstream or download speed is the term given to the data that your computer receives from the

Internet. Whenever you download music, pictures, webpages and other documents, you are using your computer's downstream bandwidth. Cable modems are typically rated at 1 to 2 megabits per second (Mbps). This means that you can download up to 2 megabits of data each second operating at full speed.

Upstream Vs. Downstream

Cable modems have what is known as asymmetrical speeds because upstream and downstream capabilities are different. Your computer is capable of downloading data two to three times faster than it is capable of uploading the same data. As the majority of functions utilize your downstream bandwidth (web browsing, downloading music, watching videos), having a significantly slower upstream is a non-issue for most people. There are a variety of "bandwidth speed tests" available on the Internet that test and display the actual speed of your cable modem.

Unicast, Broadcast and Multicast

Unicast

Traffic, many streams of IP packets that move across networks flow from a single point, such as a website server, to a single endpoint such as a client PC. This is the most common form of information transference on networks.

Broadcast

Here, traffic streams from a single point to all possible endpoints within reach on the network, which is generally a LAN. This is the easiest technique to ensure traffic reaches to its destinations. This mode is mainly utilized by television networks for video and audio distribution. Even if the television network is a cable television (CATV) system, the source signal reaches to all possible destinations, which is the key reason that some channels' content is scrambled. Broadcasting is not practicable on the public Internet due to the massive amount of unnecessary data that would continually reach at each user's device, the complications and impact of scrambling and related privacy issues.

Multicast

In this method traffic reclines between the boundaries of unicast (one point to one destination) and broadcast (one point to all destinations). And multicast is a "one source to many destinations" way of traffic distribution, means that only the destinations that openly point to their requisite to accept the data from a specific source to receive the traffic stream.

On an IP network, destinations (i.e. clients) do not regularly communicate straight to sources (i.e. servers), because the routers between source and destination must be able to regulate the topology of the network from unicast or multicast side to avoid disordered routing traffic.

Multicast routers replicate packets received on one input interface and send the replicas out on multiple output interfaces.

In the multicast model, the source and destinations are almost each time "Host" and not "Routers". The multicast traffic is spread by multicast routers across the network from source to destinations. The multicast routers must find multicast sources on the network, send out copies of packets on a number of interfaces, avoid loops, connect interested destinations with accurate source and keep the flow of unsolicited packets to a minimum. The standard protocols of multicast routing provide most of these facilities, but some router architecture cannot send multiple copies of packets and so do not support direct multicasting.

Difference between Multicast and Unicast

There are two central methods that Windows Media servers use to send data to Windows Media Player clients i.e. Unicast and Multicast. Multicast or Unicast both can be used for broadcasting live video or audio. Your network setting by default determines who your clients are and what sort of broadcast you need to prefer.

Unicast

- Traffic is sent from one host to another. A replica of each packet in the data stream goes to every host that requests it.
- The implementation of unicast applications is a bit easy as they use well-established IP protocols; however, they are particularly incompetent when there is a need for many-to-many communications. In the meantime, all packets in the data stream must be sent to every host requesting access to the data stream. However, this type of transmission is ineffective in terms of both network and server resource as it equally presents obvious scalability issues.
- This is a one-to-one connection between the client and the server. Unicast uses IP provision techniques such as TCP (transmission control protocol) and UDP (user datagram protocol), which are session-based protocols. Once a Windows media player client connects via unicast to a Windows media server that client gets a straight connection to the server. Every unicast client that connects to the server takes up extra bandwidth. For instance, if you have 10 clients all performing 100 Kbps (kilobits per second) streams, it means those clients taking up 1,000 Kbps. But you a have single client using the 100 Kbps stream, only 100 Kbps is being used.

Multicast

Multicast lets server's direct single copies of data streams that are then simulated and routed to hosts that request it. Hence, rather than sending thousands of copies of a streaming event, the server instead streams a single flow that is then directed by routers on the network to the hosts that have specified that they need to get the stream. This removes the requirement to send redundant traffic over the network and also be likely to reduce CPU load on systems, which are not using the multicast system, yielding important enhancement to efficiency for both server and network.

Is Multicast True Broadcast?

The multicast source depends on multicast-enabled routers to forward the packets to all clients' subnets that have clients listening. However, there is no direct affiliation between clients and Windows media server. The Windows media server creates an ".nsc" (NetShow channel) file when the multicast station is first formed. Usually, the.nsc file is sent to the client from a web server. This file holds data that the Windows media player requires to listen for the multicast. This is quite same to fine-tuning a station on a radio. Every client which eavesdrops to the multicast includes no extra overhead on the server. In fact, the server sends out only single stream per multicast station. The equal load is experienced on the server whether only a single client or multiple clients are listening.

Multicast on the Internet is usually not a concrete solution because only small sections of the Internet are enabled with Multicast. On the other hand, in corporate environments where all routers are multicast-enabled can save quite a bit of bandwidth.

Difference between Unicast, Broadcast and Multicast

Data is transported over a network by three simple methods i.e. Unicast, Broadcast, and Multicast. The difference between these three:

- Unicast: From one source to one destination i.e. One-to-One.
- Broadcast: From one source to all possible destinations i.e. One-to-All.
- Multicast: From one source to multiple destinations stating an interest in receiving the traffic i.e. One-to-Many.

There is no separate classification for Many-to-Many applications, for example, video conferencing or online gaming, where multiple sources for the same receiver and where receivers often are double as sources. This service model works on the basis of one-to-many multicast and for that reason requires no unique protocol. The original multicast design i.e. RFC 1112, supports both the ASM (any-source-multicast) based on many-to-many service model and the SSM (source specific multicast) based on a one-to-many model.

Digital Channel Models

Binary Symmetric Channel (BSC)

A binary symmetric channel (or BSC) is a common communications channel model used in coding theory and information theory. In this model, a transmitter wishes to send a bit (a zero or a one), and the receiver receives a bit. It is assumed that the bit is usually transmitted correctly, but that it will be "flipped" with a small probability (the "crossover probability"). This channel is used frequently in information theory because it is one of the simplest channels to analyze.

The BSC (p) has some parameter where $0 \leq p < \frac{1}{2}$. What this channel does in communication is that it sends messages in bit X, what it receives is X with probability 1 – p and 1 – X with probability p. And this happens independently on every use of the channel.

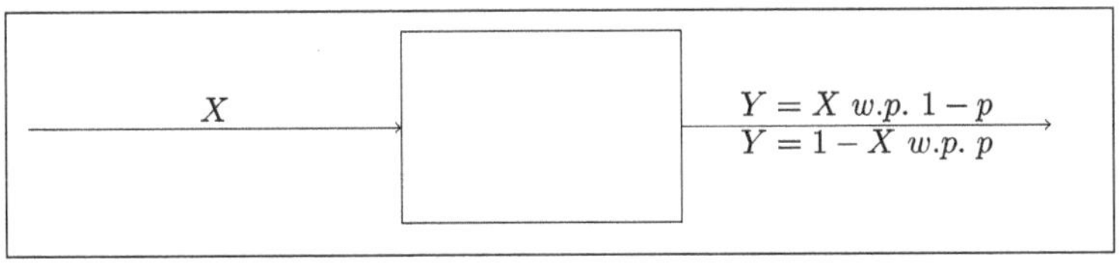

In order to use such a channel, Shannon's idea was to encode the information before you send it and decode it afterwards. Suppose you have a message m that you want to send, you encode it and get some sequence X, maybe n such symbol X, get Y and then you want to decode.

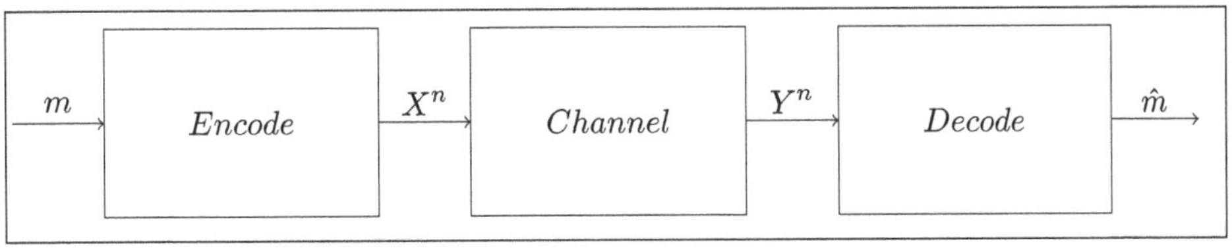

And what we want is,

$$\Pr\left[m \neq \hat{m}\right] \to 0, \text{ as } n \to \infty$$

Intuitively, if you want to send longer and longer message, the error probability should be increasing. But here there is one thing that we want to use length of message to improve the reliability. And now we would like to understand how many uses of the channel do we have to do? What is the largest R?

The encoding sends k bits to n bits, our message would be some k bit string and encoding turns it to n bit string. The decoding takes n bits and brings it back down to k bits.

$$E_n : \{0, 1\}^k \to \{0, 1\}^n, \; D_n : \{0, 1\}^n \to \{0, 1\}^k$$

Where, $k \leq n$

Example: The rate would be Rate = $\frac{k}{n}$.

What we would really want to understand is the capacity of the channel. In this case the capacity of the channel would be,

$$\text{Capacity}(\text{BSC}(p)) = \sup_R \left\{ \lim_{\epsilon \to 0} \lim_{n \to \infty} \left\{ \text{Rate R of } E_n, D_n \right\} \right\}$$

Such that,

$$\Pr_{m, y|x, X=E(m)} \left[D(y^n) \neq m \right] \leq \epsilon$$

Allow to use n as large as you get but have to make sure that error goes to zero when n goes to infinity. This is the general quantity we want to understand for any channel.

Remark: The capacity of binary symmetric channel is,

$\text{Capacity}\left(\text{BSC}_{(p)}\right) = 1 - h(p)$ where $h(p) = p\log\frac{1}{p} + (1-p)\log\frac{1}{1-p}$ which the entropy of Bernoulli p random variable is.

This is a little striking theorem. Why we get the 1 – h(p)? Roughly the idea is the following, X came in the channel with a sequence of Bernoulli random variables η distributed according to Bern(p), after the decoding, suppose you are able to reproduce the original message m, then you can easily use this to determine also η. This channel is sending for free after the decoding n Bernoulli random bits that should require nh(p) uses of the channel. What is remaining to use is 1 – h(p) and that's what we're going to use to convey X.

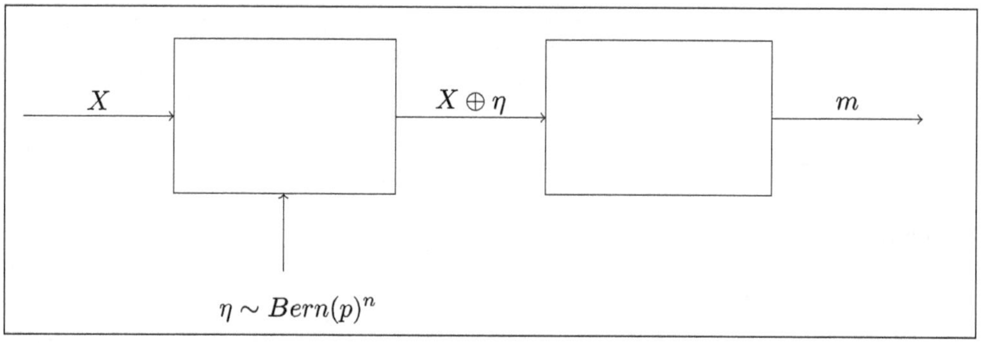

If you try to draw the capacity of the channel 1 – h(p) and h(p). When p = 0 you get capacity 1 which means that for every use of the channel you can send 1 bit through which makes sense. When p = 1 2 you get no capacity which also makes sense when you send 0 or 1 the receiver receives unbiased bit so there is no correlation between sending and receiving.

General Channel

It turns out when you are talking about general channel; you get a even nicer connection. We would like to find the rate and capacity for the general channel. First we would like to specify the encoder for general channel. For today general channel means memory less and here we're just taking some arbitrary channel which acts independently on every single bit being transmitted.

The input X is some element of some universe Ω_x. The output some element of some universe Ω_y. The universe means not to be related. We want think about stochastic channels which make a lot sense that channel are given by a bunch of conditional distributions.

Example: One very simple example of this may be called erasure channel. The following figure shows a Binary Erasure Channel (BEC). It produces a output of 0 or 1 or?

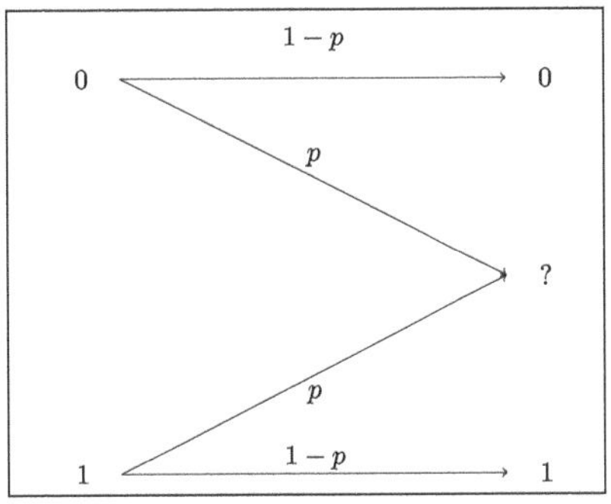

Example: People in information theory may also think about something called noisy typewriter. The channel input is either unchanged with probability $\frac{1}{2}$ or transformed into the next letter with probability $\frac{1}{2}$ in the output.

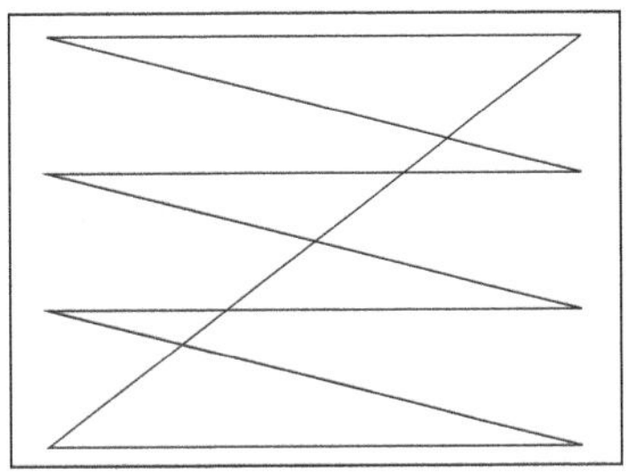

Remark: The matrix $P_{y|x}(\alpha, \beta)$ specifies the channel.

$$P_{y|x}(p \mid \alpha) = \Pr[\text{Channel outputs } \beta \in y, \text{ information input } \alpha \in \Omega_x]$$

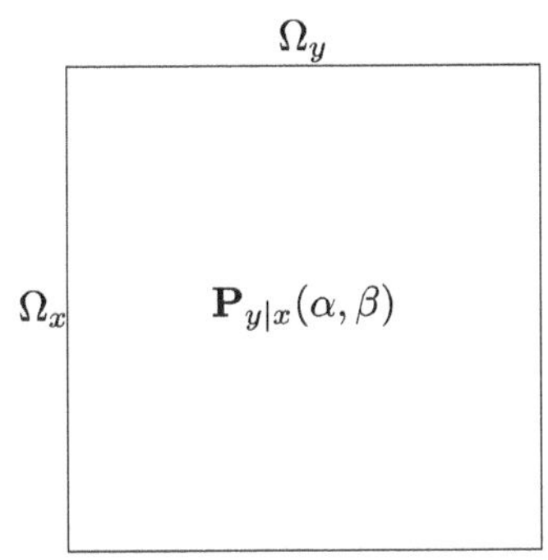

For the binary symmetry channel case, we have,

$1-p$	p
p	$1-p$

Coding Theorem

An encoding function would be a map from {0, 1} ^{n} to Ω_x^n the decoding function would be a map from Ω_y^n $\{0,1\}^k$. And rate is still defined as Rate $= \frac{k}{n}$. And we still have the probability of error and we still can talk about capacity of channel. For any general channel, the capacity is,

$$\text{Capacity}\left(\text{Channel}\, P_{y|x}\right) = \sup_R \left\{ \lim_{\epsilon \to 0} \lim_{n \to \infty} \left\{ \text{Rate R of } E_{n,}\, D_n \right\} \right\}$$

Here we will get the remarkable theorem once again. It turns out the capacity of the channel are given by this joint distribution.

Theorem:

$$\text{Capacity}\left(P_{y|x}\right) = \sup_{P_x} \; I(x; y)$$

Remark: Once X distribution is specified, I get a joint distribution on (X, Y). The information that Y conditioned on X is the capacity of this channel. This completely characterizes every memory less channel of communication. Given Y, figuring out X. Then mutual information is the right characterization.

Let's prove half of the statement first, half of that statement next, and then other half. Here is what we're going to do with the encode, pick n large enough, let k ≥ (1 − h(p) − ε)n, let $E_n : \{0, 1\}^k \to \{0, 1\}^n$ completely random. The decoding function is, not changed, given some sequence Y, we're going to look at the m which maximizes the probability y n conditioned on x^n, that is,

$$Dn\left(\beta^{\,n}\right) = \underset{m}{\operatorname{argmax}} \left\{ P_{y^n|x^n} \; (\beta^n, E(m) \right\}$$

Where, $\beta^n \in \Omega_y^n$. It could be used to talk any channel, for instance, the Markov channels. The following theorem is nor hard to prove. We have,

Theorem:

$$\Pr_{E_{n,m,y|E_n(m)}} \left[D_n(y) \neq m \right] \leq \epsilon$$

$$\Rightarrow \exists E_n \Pr_{m,y|E_n(m)} \left[D_n(y) \neq m \right] \leq \epsilon$$

Remark:

$$\Pr_{E_{n,m,y,BSC_P}} \left[D_n\left(BSC_P\left(E_n(m)\right)\right) \neq m \right] \leq \epsilon$$

$$\Rightarrow \exists E_n \Pr_{m,y,BSC_P} \left[D_n\left(BSC_P\left(E_n(m)\right)\right) \neq m \right] \leq \epsilon$$

Proof: There are two types of errors.

- Error of type 1(E_1): Too many errors happen. X^n and Y^n differ in more than (P + ε)n coordinates. It's straightforward to show that $\Pr[E_1] \leq \exp(-n)$.

- Error of type 2(E_2): Fix message m you just transmitted, fix the encoding of this message $E_n(m)$, that is an random variable but it's fixed and fix $y = BSC_p(E_n|_m)$, let $E^1_n : \{0, 1\}^k - \{m\} \rightarrow \{0, 1\}^n$ is not fixed and be random. E_2 is the event that there exists mo such that encoding of E_n (m') and y disagree in less than or equal to $(p + \varepsilon)n$ coordinates. We want neither of these events happen then the most likely message is the one we transmitted.

$$\Pr\left[D_n\left(BSC\left(E_n(m)\right)\right) \neq m\right] \leq \Pr\left[E_1 \cup E_2\right] \leq \Pr\left[E_1\right] + \Pr\left[E_2\right]$$

$P_r[E_1]$ is exponentially small thus we are left with probability E_2, the bound rely on the probability of E_2 which is simple to calculate.

Lemma:

$$\forall m', \Pr\left[m' \in \text{Ball of radius } (P+\varepsilon)n \text{ around } y\right] = \frac{\text{Volume(Ball)}}{2^n} \approx \frac{2^{(h(p)+\varepsilon)n}}{2^n}$$

Remak:

$$\text{Ball}_r(y) = \left\{Z \in \{0, 1\}^n \mid Z \text{ and y differ in} \leq r \text{ coordinates}\right\}$$

The volume of this ball or the size of this set is:

$$\left|\text{Ball}_r(y)\right| = \sum_{i=0}^{r}\binom{n}{i} \approx 2^{h\left(\frac{r}{n}\right)n}$$

That's just one single message, if you want any valid message,

$$P\,r\left[\exists m', \text{ st. } E_2\right] \leq \frac{2^k 2^{(h(p)+\varepsilon)n}}{2^n}$$

This is where we see the quantity that we want, $\frac{k}{n} + h(p) + \varepsilon < 1$ and why need one minus entropy.

Lemma: If $Z_1 \ldots Z_n$, $Z_i \sim P_z$ iid, then,

$$\exists S \subseteq \Omega_z^n \frac{1}{|S|^{1+\epsilon}} \leq \Pr\left[Z_1 \ldots Z_n = r_1 \ldots r_n\right] \leq \frac{1}{|S|^{1-\epsilon}}$$

$$\Pr\left[(Z_1 \ldots Z_n) \notin S\right] \leq \varepsilon$$

Here S="typical set for P_z"

We are going to do the usual three step analysis. The first step is to pick the encoding of the message E(m), the second step is to pick what you received conditioned on what you send y|E(m) and the third step is to look at E(m') where m' ≠ m. There are three kinds of typicality errors. If x = E (m) is not be typical for P_x, we say we have error of type 1. y may not be typical for P_y. The most important thing is we also want (X, Y) to be typical. But (X, Y) may not be typical for P_{xy}. Probability of any of these errors happens is negligible by AEP. But we have to get the mutual information out of these.

Remark: The decoding is some function of β and if β is not typical for P_y, then declare an error. If there exists a unique m such that E(m) is typical for Px and (E(m), y) are typical for P_{xy}, then output m.

We are going to look at E(m') and want to analyze the probability that E(m') is typical and E(m', y) is also typical. Let's fix the m' and ask the question. The fact E (m') is typical is going to happen in a very high probability.

We are more interested in the probability that (E(m'), y) is jointly typical. This is the crucial question. This is actually distributed according to $P_x \times P_y$. Here is the lemma that will state which immediately imply.

Lemma: Let $Z_n \sim P^n$,

$$\Pr\left[Z \text{ is typical for some distribution } Q\right] \leq 2^{D(Q\|P)}$$

This is another fundamental reason to understand the divergence between these two distributions. You can apply to very simple things. In the current case we are looking into $D(P_{xy}||P_x \times P_y) = I(X, Y)$. When you combine these two facts, apply here, it turns out that any particular message is going to be jointly typical, the probability is approximately two to the minus mutual information. It tells the rate is at least the mutual information for this distribution.

Binary Erasure Channel (BEC)

The binary erasure channel (denoted by BEC (ε)) is one of the simplest examples of binary memory less symmetric channel (BMS). As explained by the name, these are binary channel in the sense that the input is a bit X ∈ {0, 1}. Any input X leads to an output Y ∈ {0, 1, ∗} which is distributed according to:

$$Y = \begin{cases} X & \text{with probability } 1 - \varepsilon, \\ * & \text{with probability } \varepsilon, \end{cases}$$

independently across channel uses. In other words, each input symbol is erased independently with probability ε ∈ [0, 1].

A channel code with blocklength n and rate R ∈ (0, 1] is a mapping from binary strings of length nR, to binary strings of length n,

$$\text{Enc}: \{0, 1\}^{nR} \rightarrow \{0, 1\}^n$$

Sometimes a code is defined as a subset $\mathfrak{C} \subseteq \{0,1\}^n$ with cardinality $|\mathfrak{C}| = 2^{nR}$. This is nothing but the image of Enc: $\mathfrak{C} = \text{Enc}\left(\{0, 1\}^{nR}\right)$ The two definitions are nearly equivalent and we will use use either one depending on what is most convenient.

If the channel output belongs to an alphabet $\mathcal{Y}$, a decoder can be defined aDec: $\mathcal{Y}^n \rightarrow \{0, 1\}^{nR}$ (recovering the input of the encoder) or as a mapping:

$$\text{Dec}: \mathcal{Y}^n \rightarrow \{0,1\}^n$$

(Recovering the channel input). The two definitions are essentially equivalent as long as Dec(y) ∈ $\mathfrak{C}$ with high probability, but we will generally adopt the latter since it is more convenient for our analysis. We will also denote the decoder by $\hat{x} : \mathcal{Y}^n \to \{0, 1\}^n$.

Given a code $\mathfrak{C}$ and a communication channel (for instance BEC(ε)), two standard figures of merit are the block error rate and the bit error rate,

$$P_B(\mathfrak{C}, \varepsilon) = \frac{1}{|\mathfrak{C}|} \sum_{x \in \mathfrak{C}} \mathbb{P}\,(\hat{x}(y) \neq x),$$

$$P_b(\mathfrak{C}, \varepsilon) = \frac{1}{|\mathfrak{C}|} \sum_{x \in \mathfrak{C}} \sum_{i=1}^{n} \mathbb{P}\,(\hat{x}(y)_i \neq x_i),$$

Note that these quantities depend on the decoding algorithm as well. We will leave this dependence implicit unless needed.

We say that a sequence of codes $(\mathfrak{C}_n)_{n \geq 1}$ of rate R indexed by the block length n (and associatex decoders $\hat{x}(\cdot)$) allows to communicate reliably over the channel BEC (ε) if $\lim_{n \to \infty} PB(\mathfrak{C}_n, \varepsilon) = 0$.The same definition applies unchanged to other channel models.

Shannon's channel coding theorem established that reliable communication is possible if and only if the rate is smaller than the channel capacity. In the case of the BEC (ε), this theorem takes an extremely simple form. Reliable communication is possible if and only if,

$$R < C_{BEC}(\varepsilon) = 1 - \varepsilon$$

It is possible give an elementary proof that reliable communication is impossible for R > 1 – ε. Indeed for a given output $y \in \{0, 1, *\}^n$, let $S \subseteq [n]$ be the subset of bits that is not erased in y. For a given S, the decoder will take at most $2^{|S|}$ values (corresponding to the possible values of x_S), and hence it will be incorrect on at least $2^{nR} - 2^{|S|}$ input codewords (for $|S| \leq nR$). We therefore have,

$$P_B(\mathfrak{C}, \varepsilon) \geq \mathbb{E}\{1 - 2^{-nR+|S|}\}\,\mathbb{I}(|S| \leq nR)\}$$

Note that $|S| \sim \text{Binom}(n, 1 - \varepsilon)$, and hence we have $|S| \approx n(1 - \varepsilon)$ with high probability for large n. For $R > C_{BEC}(\varepsilon)$, we have $|S| \leq nR - 1$ with probability bounded away from 0, and hence PB(C, ε) stays bounded away from 0.

Dense Graphs

A linear code is defined by a n(1 – R) × n binary matrix $\mathbb{H}$. In the following we set m = n(1 – R). The codebook is,

$$\mathfrak{C} = \{x \in \{0, 1\}^n : \mathbb{H}x = 0 \mod 2\}$$

Such a code can also be completely described by a bipartite graph, namely the graph whose adjacency matrix is $\mathbb{H}$.

The parity check ensemble is a randomized construction of the matrix H, first introduced by Peter Elias. It consists in taking each entry H_{ij} to be 0 or 1 independently of the others with probability 1/2. We might also call this the high-density parity check (HDPC) ensemble. Here the density is defined as the fraction of ones in the parity check matrix. An example of a graph from this ensemble (with blocklength n = 16 and design rate R = 1/4) is below.

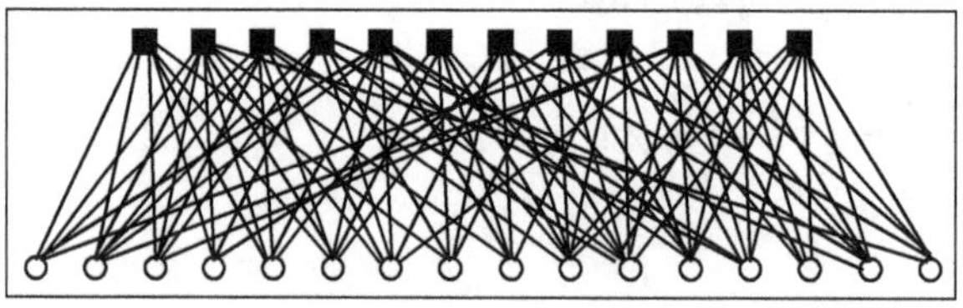

This ensemble can be used to communicate over the BEC(ε) at vanishing error probability if R < $C_{BEC}(\varepsilon) = 1 - R$. In order to prove this claim, denote again by $S \subseteq [n]$ the subset of position that is not erased by the channel, and let $S^c = [n]\backslash S$. We let $\mathbb{H}_A$ be the sub matrix of H formed by the columns with index in A. We can decode the message y by solving the following linear system in the unknown x_{Sc}:

$$\mathbb{H}_S x_S + \mathbb{H}_{S^c} x_{S^c} = 0.$$

This has a unique solution (which necessarily coincides with the transmitted codeword) if and only if H_{Sc} has rank equal to $|S^c| = k$. Of course this is possible only if $k \le m \equiv n(1 - R)$. We denote by $P_{m,k}$ the probability that a uniformly random m × k matrix with entries in {0, 1} has rank equal to k. Note that $P_{m,1} = 1 - 2^{-m}$ because in this case the rank is 1 except if the matrix is all zeros. Further, we have the recursion, for k ≥ 2,

$$P_{m,k} = P_{m,k-1}\left(1 - \frac{2^{k-1}}{2^m}\right)$$

Indeed, for the rank to be k, the first k – 1 columns must be full rank, and the last column must not be in the span of the first k – 1 ones. Hence we have,

$$P_{m,k} \prod_{l=0}^{k-1}(1-2^{-m+l})\mathbb{I}(m \ge k)$$

The expected block error probability over this code ensemble reads,

$$\overline{P_B}(\varepsilon) = \mathbb{E}\{P_B(\mathfrak{C}_n;\varepsilon)\} = 1 - \sum_{k=0}^{m} \mathbb{P}(|S| = k)\, P_{m,k}$$

$$1 - \sum_{k=0}^{m} \binom{n}{k} \in^k (1-\in)^{n-k} P_{m,k}$$

In order to estimate this probability, we note that, with high probability |S| ≤ nε' for any ε' > ε. Taking ε' < (1 – R), and using the fact that Pm,k is decreasing in k, and assuming for simplicity ε' an integer, $\overline{P_B}(\varepsilon) \le 1 - P_{m,n\varepsilon'} + o_n$

$$= 1 - \exp\left\{ \sum_{l=0}^{n\varepsilon'-1} \log(1 - 2^{-m+l}) \right\} + o_n(1) = 1 - \exp\left\{ -n\varepsilon' 2^{-n(1-R+\varepsilon')} \right\} + o_n(1) = o_n(1)$$

We thus proved that the random linear code allows achieving capacity over the erasure channel. Decoding amounts to solving a linear system of size $n(1 - R) \times n\varepsilon$, which can be done in time $O(n^2)$ via Gaussian elimination. A simpler algorithm is provided by the following greedy decoder, which we shall call peeling algorithm.

Peeling Decoder (A partially erased codeword y):

1: Set $\underline{x} = \underline{y}$;

2: Set `flag = false`

3: For $a \in C$

4: If ∂a includes exactly one erased position

5: Set its value to the sum of the other variables in ∂a

6: Set `flag = true`

7: End for;

8: If `flag = true`

9: Goto 1;

10: Else

11: Return current guess $\underline{x}$

Unfortunately, the peeling decoder performs quite poorly on HDPC ensemble. The reason is that it is quite unlikely to find a parity check that contains only one erased position.

Sparse Graphs

A good idea is to make the factor graph sparse. We will say that the graph is sparse if the fraction of ones in the parity check matrix is O (1/n). Equivalently the average degree of the graph is O (1).

A simple random ensemble can be defined in this way: for each variable node $i \in [n]$ and each check node $a \in [m]$, draw an edge (i, a) independently with probability γ/n. Here γ is a density parameter that is kept fixed as $n \to \infty$. An example is drawn below.

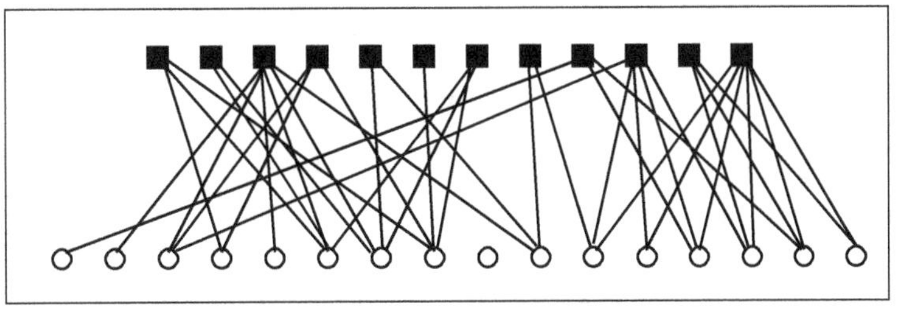

Almost One-dimensional Graphs

Let's try something different (here explained for the case R = 1/4). Arrange variable nodes and check nodes on a line of length 3n (= 4m) and connect each check node to the 4 variable nodes that follow it on the line. When you arrive at the end of the line, 'wrap it around.' In the figure below is sketched this construction but placed nodes on two rows to reduce the messiness.

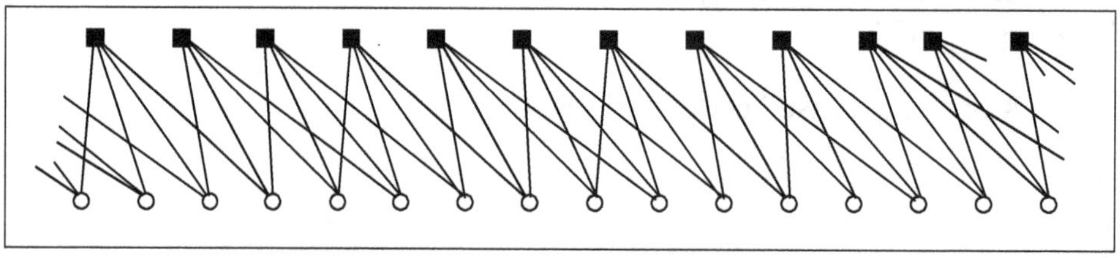

Regular Graphs

In regular ensemble the factor graph is random with degree l at variable nodes and k at check nodes. The (design) rate is R = 1–l/k. We also refer to this at the (l, k) ensemble. Ideally the graph would be uniformly random.

In practice one uses a modified ensemble defined by the following generation procedure. First draw 3 half-edges for each variable node and 4 half-edges for each check node. Then draw a uniformly random permutation over 3n elements (how would you do this?). Finally connect half edges on the two sides according to this permutation. The resulting graph might have double edges.

Message Passing

Consider a code with factor graph G. Each node (either check or variable) is a little processing unit (I usually imagine a little genie inside the node.) Nodes exchange messages along edges, with the objective of decoding the transmitted codeword. 'Messages' are elements of an alphabet $\mathcal{M}$ that you are free to choose at your convenience.

At any given time, a node can only use the messages it received so far, and (if it is the variable node) the channel output at its position. For the sake of simplicity, we'll assume that nodes are perfectly synchronized: at each iteration all variable nodes send out messages, and then all check nodes, and so on.

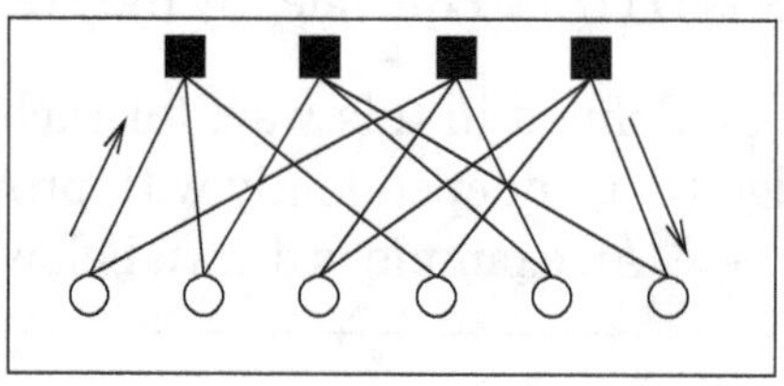

Consider now decoding a message transmitted over the BEC (ε):

- Assume the degree of G is bounded by d, and the alphabet size $|\mathcal{M}|$ to be finite. How many operations does it take to run t iterations?
- Is there a way to implement the peeling decoder in the above model?

- Consider now a further restriction. The message sent at any time from node i to node a only depends on messages reaching i at the previous iteration from other nodes. Is it still possible to implement the peeling decoder?

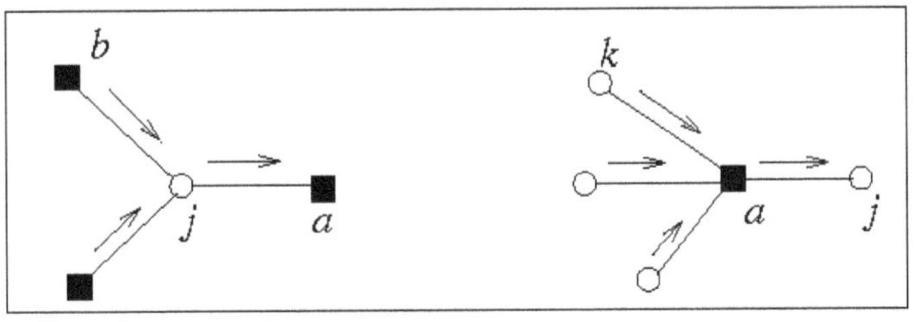

- In the previous, what would you use as a stopping criterion? After how much iteration will the algorithm stop under this criterion?
- Suppose you had to implement a message passing procedure on your computer. Which data structure would you use to store the graph G?

Density Evolution

Let $x_t \in [0, 1]$ be the probability that a uniformly random message from variable to check nodes is wrong after t iterations. In the limit $n \to \infty$, this satisfies the recursive equation,

$$x_{t+1} = \varepsilon\left(1 - (1-x_t)^{k-1}\right)^{l-1},$$

With initial condition $x_0 = \varepsilon$. What can you say about this recursion? Let $\lambda(z) \equiv z^{l-1}$ and $\rho(z) = z^{k-1}$. Then the recursion can be written as,

$$x_{t+1} = \varepsilon\lambda\left(1-\rho\left(1-x_t\right)\right),$$

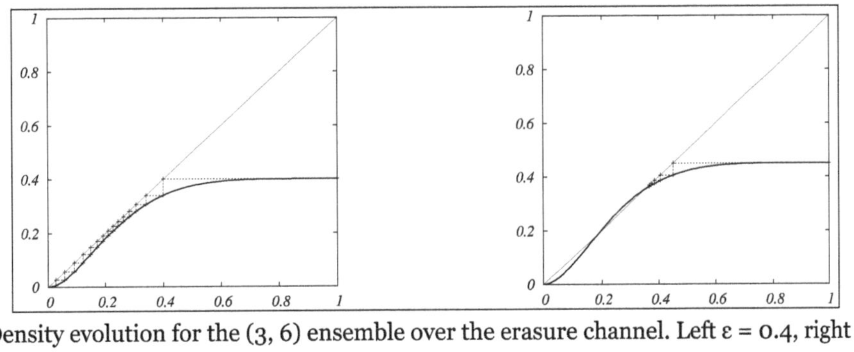

Figure: Density evolution for the (3, 6) ensemble over the erasure channel. Left ε = 0.4, right ε = 0.45.

Or,

$$\lambda^{-1}\left(x_{t+1}/\varepsilon\right) = 1-\rho\left(1-x_t\right)$$

Represent this last form of the recursion graphically. Notice that,

$$\int_0^1 \varepsilon\lambda(x)\,dx = \frac{\varepsilon}{l}, \int_0^1 \rho(x)\,dx = \frac{1}{k}$$

Interpret these identities graphically.

Analog Channel Models

Noise Model

Noise can be defined as any unwanted signals, random or deterministic, which interfere with the faithful reproduction of the desired signal in a system. Stated another way, any interfering signal, which is usually noticed as random fluctuations in voltage or current tending to obscure and mask the desired signals is known as noise. These unwanted signals arise from a variety of sources and can be classified as man-made or naturally occurring.

Man-made types of interference (noise) can practically arise from any piece of electrical or electronic equipment and include such things as electromagnetic pick-up of other radiating signals, inadequate power supply filtering or alias terms. Man-made sources of noise all have the common property that their effect can be eliminated or at least minimized by careful engineering design and practice.

Interference caused by naturally occurring noise is not controllable in such a direct way and their characteristics can best be described statistically. Natural noise comes from random thermal motion of electrons, atmospheric absorption and cosmic sources.

Statistical Description of Signals

Since noise is mostly random in nature, it is best described through its statistical properties. Here, the main parameters and in-between relations for noise description are presented and analysed. Without going into details of random variables and stochastic processes, expressions are developed to describe noise through its power spectral density (frequency domain) or equivalently the auto-correlation function (time domain).

Time-averaged Noise Representations

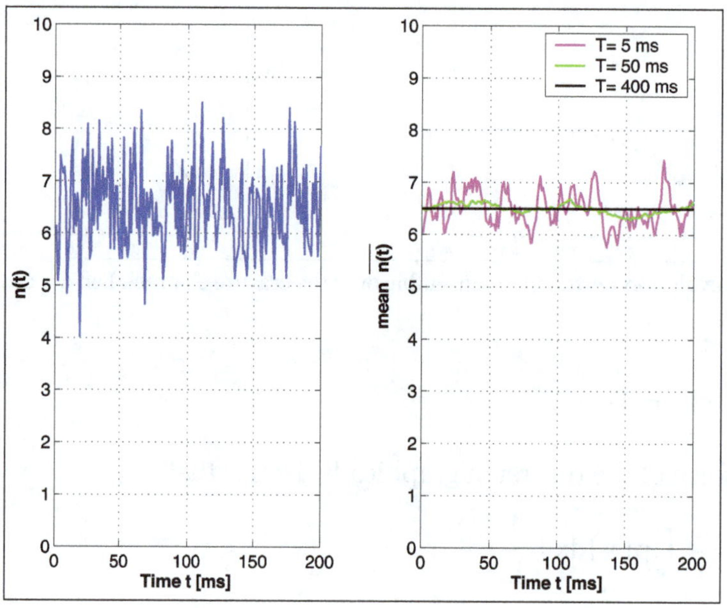

Figure: A random noise waveform and its average value.

In forming averages of any signal, whether random or deterministic, we find parameters which tell us something about the signal. But much of the detailed information about the signal is lost through this process. In the case of random noise, however, this is the only useful quantity.

Suppose n(t) is a random noise voltage or current. A typical waveform is illustrated in figure. We now define the following statistical quantities of n (t).

Mean Value

The mean value of n (t) will be referred to as $\overline{n(t)}$, η_T or E {n (t)}. It is given by:

$$\overline{n(t)} = E\{n(t)\} = \eta_T = \lim_{T\to\infty}\frac{1}{T}\int_{-T/2}^{T/2} n(t)\,dt$$

Where $\overline{n(t)}$ is often referred to as dc or average value of n(t). For practical calculation of the mean value, the averaging time T has to be chosen large enough to smooth the fluctuations of n(t) adequately. Figure shows the averages $\overline{n(t)}$ calculated by sliding a window centered at t and extending from t – T/2 to t + T/2 over n (t). It is seen that for small averaging time T = 5 ms there is still a considerable amount of fluctuation present whereas for T = 400 ms the window covers the whole time signal which results in a constant average value.

Mean-square Value

$$\overline{n^2(t)} = E\{n^2(t)\} = \lim_{T\to\infty}\frac{1}{T}\int_{-T/2}^{T/2} |n(t)|^2\,dt$$

Aside from a scaling factor the mean-square value $\overline{n^2(t)}$ in $\overline{n^2(t)} = E\{n^2(t)\} = \lim_{T\to\infty}\frac{1}{T}\int_{-T/2}^{T/2} |n(t)|^2\,dt$ gives the time averaged power P of n(t). Assuming n(t) to be the noise voltage or current, the scaling factor will be equivalent to a resistance, which is often set equal to 1Ω. The square root of $\overline{n^2(t)}$ is known as the root-mean-square (rms) value of n(t). The advantage of the rms notation is that the units of $\sqrt{\overline{n^2(t)}}$ are the same as those for n(t).

AC Component

The ac or fluctuation component σ (t) of n (t) is that component that remains after "removing" the mean value $\overline{n(t)}$ and is defined as:

$$\sigma(t) = n(t) - \overline{n(t)}$$

Variance The variance $\overline{\sigma^2(t)}$ defined by:

$$\overline{\sigma^2(t)} = E\{\sigma^2(t)\} = E\left\{\left(n(t) - \overline{n(t)}\right)^2\right\} = \lim_{T\to\infty}\frac{1}{T}\int_{-T/2}^{T/2} |\sigma(t)|^2\,dt,$$

is equal to the power of the ac component of n(t) (aside from a resistance scaling factor). This can be showed by substituting $\sigma(t) = n(t) - \overline{n(t)}$. into $\overline{n^2(t)} = E\{n^2(t)\} = \lim_{T\to\infty} \frac{1}{T}\int_{-T/2}^{T/2} |n(t)|^2 \, dt$ giving:

$$\overline{n^2(t)} = \lim_{T\to\infty} \frac{1}{T}\int_{-T/2}^{T/2} \left|\overline{n(t)} + \sigma(t)\right|^2 dt$$

Using the fact that $\overline{n(t)}$ is a constant and that the mean of σ(t) is zero by definition, we get:

$$\overline{n^2(t)} = \left|\overline{n(t)}\right|^2 + \lim_{T\to\infty} \frac{1}{T}\int_{-T/2}^{T/2} \left|\sigma(t)^2\right| dt$$

In the above equation, the time averaged power of n (t) is written as the power sum of the dc and ac signal components. The variance is a measure of how strong the signal is fluctuation about the mean value.

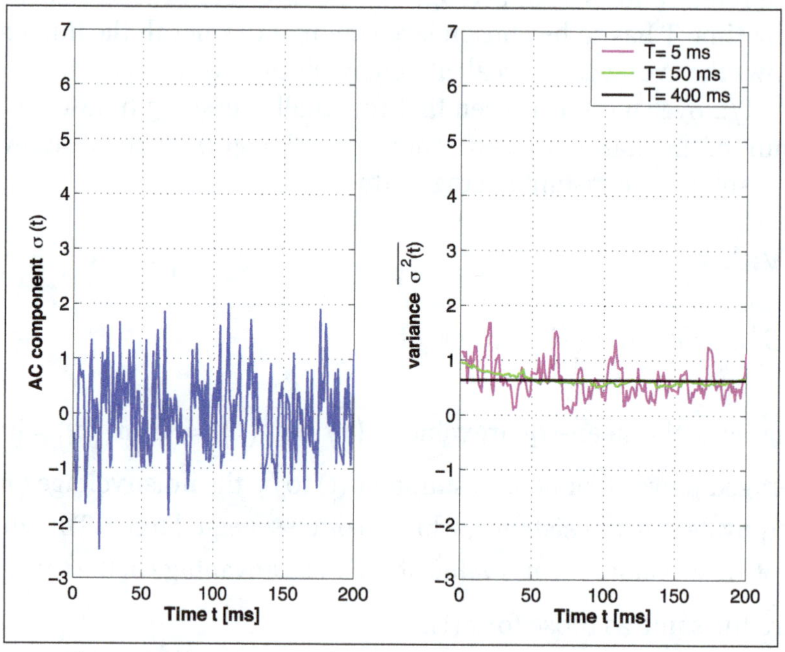

Figure: AC component and variance of random noise waveform.

Validity of Time-based Statistics

Mathematically noise is modeled as a stochastic process, where the noise can be considered as the realization of a random variable. Consider the stochastic process x(t), at time instance t_1, $x(t_1)$ is a random variable; its expected value (mean) is $\eta(t_1) = E\{x(t_1)\}$, and can be estimated by observing N samples (i.e. realizations) $x(t_1, \xi_i)$ through:

$$\hat{\eta}(t_1) = \frac{1}{N} \sum_{I} x(t_1, \xi_i)$$

Where, it can be shown that $\hat{\eta}(t_1)$ is a consistent estimator of $\eta(t_1)$. However, the above expression can only be used if a large number of realizations $x(t_1, \xi_i)$ of $x(t_1)$ are available. In many applications this is not the case and mostly only a single sample (realization) of $x(t, \xi_1)$ is available. We therefore estimate $\eta(t)$ in terms of the time average of the given sample. For this to be possible, first, $E\{x(t)\}$ should not depend on t; in this case x(t) is called a stationary process. To arrive at the second condition, we form the time average according to:

$$\eta_T = \frac{1}{T}\int_{-T/2}^{T/2} x(t)dt$$

Which is a random variable with mean:

$$E\{\eta_T\} = \frac{1}{T}\int_{-T/2}^{T/2} E\{x(t)\} = \eta$$

Now if the time average η_T computed from a single realization of x(t) tends to the ensemble average η as $T \to \infty$ then the random process is called mean-ergodic. As such, for ergodic random processes a single time realization can be used to obtain the moments of the process. Thus, the expressions in previous equations are only correct if the stochastic process is both stationary and ergodic. In addition, since the measuring time T is finite the quantities are only estimated values of the moments. In practice, the only quantity accessible to measurements is n(t), which forces us to assume a stationary, ergodic stochastic process.

Fourier Transform

The definition of the Fourier transform is given by,

$$F(f) = \mathcal{F}\{f(t)\} = \int_{-\infty}^{\infty} f(t)e^{-j2\pi ft}dt$$

And the inverse Fourier transform,

$$f(t) = \mathcal{F}^{-1}\{F(f)\} = \int_{-\infty}^{\infty} F(f)e^{j2\pi ft}df$$

If the signal f (t) is a power signal, i.e., a signal with finite power but infinite energy, the integral in $F(f) = \mathcal{F}\{f(t)\} = \int_{-\infty}^{\infty} f(t)e^{-j2\pi ft}dt$ will diverge. However, considering the practical case of a finite observation time T and assume that the signal is zero outside this interval, is equivalent to multiplying the signal by the unit gate function rect (t/T). In this case the Fourier transform can be written as:

$$F_T(f) = \mathcal{F}\{f(t)\text{rect}(t/T)\} = \int_{-T/2}^{T/2} f(t)e^{-j2\pi ft}dt$$

Note that the multiplication by the rect-function in the time domain is equivalent to a convolution by a sin c-function in the frequency domain.

Correlation Functions

In the following, two statistical functions are introduced which can be used to investigate the similarity between random functions.

Auto-correlation Function

The auto-correlation function $E\{f*(t)f(t+\tau)\} = R_f(\tau)$ of a signal f(t) is defined as:

$$E\{f^*(t)f(t+\tau)\} = R_f(\tau) = \lim_{T\to\infty}\frac{1}{T}\int_{-T/2}^{T/2} f^*(t)f(t+\tau)dt$$

Where, $f^*(t)$ is the complex conjugate of f(t). Note that the subscript f is added to the autocorrelation function $R(\tau)$ to indicate the random variable or function that is considered. The auto-correlation function $E\{f^*(t)f(t+\tau)\} = R_f(\tau) = \lim_{T\to\infty}\frac{1}{T}\int_{-T/2}^{T/2} f^*(t)f(t+\tau)dt$ is often used in signal analysis, it gives a similarity measure of the signal f(t) with itself versus a relative time shift by an amount τ. For slowly varying time signals, the signal values don't change rapidly over time which will result in a flat auto-correlation function $R_f(\tau)$. Noise signals on the other hand, tend to have rapid fluctuations giving rise to an auto-correlation function with a sharp peak for $\tau = 0$ (no time shift) and quickly falling to zero for increasing τ. As an example Figures, show the time signals and the corresponding autocorrelation functions for both an exponential and a random noise signal.

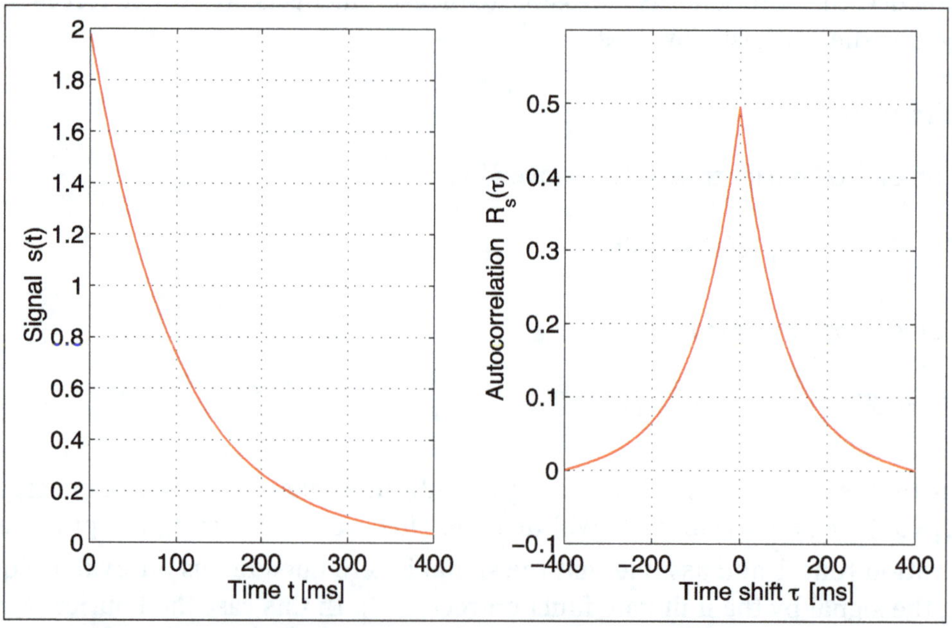

Figure: Time signal and auto-correlation function of exponential waveform.

When dealing with random variables, the auto-correlation function $R_f(\tau)$ is a statistical quantity describing the stochastic process.

The setting $\tau = 0$ in $E\{f^*(t)f(t+\tau)\} = R_f(\tau) = \lim_{T\to\infty}\frac{1}{T}\int_{-T/2}^{T/2} f^*(t)f(t+\tau)dt$

yields $R_f(0) = E\{f^2(t)\} = \overline{f^2(t)}$ which is the average power of the signal as is readily seen by comparing to $\overline{n^2(t)} = E\{n^2(t)\} = \lim_{T\to\infty}\frac{1}{T}\int_{-T/2}^{T/2}|n(t)|^2 dt$. Again, we note that the definition of $R_f(\tau)$ as given by $E\{f^*(t)f(t+\tau)\} = R_f(\tau) = \lim_{T\to\infty}\frac{1}{T}\int_{-T/2}^{T/2} f^*(t)f(t+\tau)dt$ is only valid if the stochastic process is both stationary and ergodic. It can be shown, that taking the Fourier transform (with respect to τ) of both sides of $E\{f^*(t)f(t+\tau)\}=R_f(\tau)=\lim_{T\to\infty}\frac{1}{T}\int_{-T/2}^{T/2}f^*(t)f(t+\tau)dt$ yields:

$$\mathcal{F}\tau\{R_f(\tau)\}=\lim_{T\to\infty}\frac{1}{T}F_T^*(f)F_T(f)=\lim_{T\to\infty}\frac{1}{T}|F_T(f)|^2$$

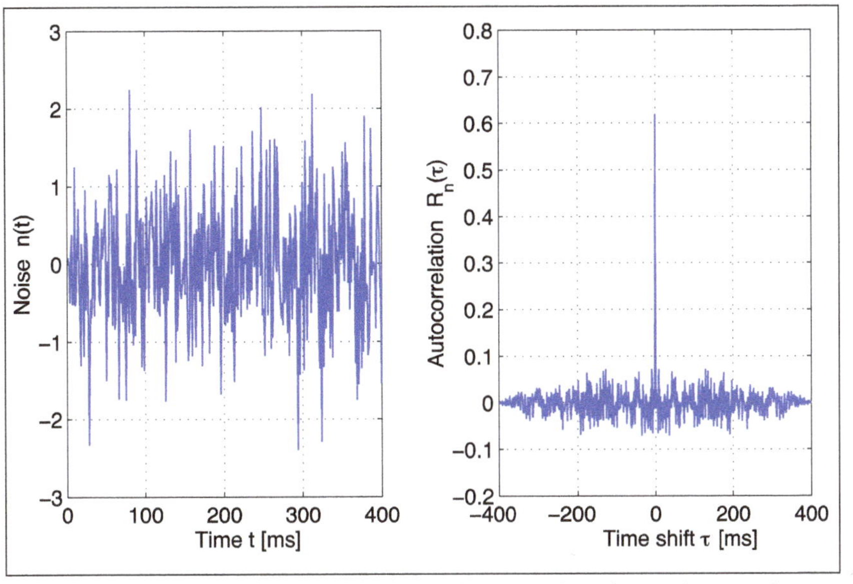

Figure: Time signal and auto-correlation function of random noise waveform.

Thus, we arrive at the important conclusion, that the correlation integral resulting in Rf (τ) corresponds to a multiplication in frequency domain. Or, equivalently we can say that instead of evaluating the integral in $E\{f^*(t)f(t+\tau)\}=R_f(\tau)=\lim_{T\to\infty}\frac{1}{T}\int_{-T/2}^{T/2}f^*(t)f(t+\tau)dt$ we can calculate the Fourier transform of f(t) according to $F_T(f)=\mathcal{F}\{f(t)\mathrm{rect}(t/T)\}=\int_{-T/2}^{T/2}f(t)e^{-j2\pi ft}dt$, determine |FT(f)|2 and use the inverse Fourier transform to get Rf(τ). The limit for T → ∞ in $\mathcal{F}\tau\{R_f(\tau)\}=\lim_{T\to\infty}\frac{1}{T}F_T^*(f)F_T(f)=\lim_{T\to\infty}\frac{1}{T}|F_T(f)|^2$ mainly reminds us of the finite observation time, practically this limit means that we have to observe the signal for a sufficient time period.

Cross-correlation Function

A function closely related to the auto-correlation function, which can be used as a similarity measure between two different signals, is the cross-correlation function. For two waveforms f(t) and g(t), the cross-correlation function $E\{f^*(t)g(t+\tau)\} = C_{fg}(\tau)$ is defined as,

$$E\{f^*(t)g(t+\tau)\} = C_{fg}(\tau) = \lim_{T\to\infty}\frac{1}{T}\int_{-T/2}^{T/2}f^*(t)g(t+\tau)dt$$

The auto-correlation is considered a special case of the cross-correlation, since it can be obtained by setting f(t) = g(t) in previous equation. The interpretation of the cross-correlatio between two signals is similar to that of the auto-correlation, with the advantage tha additional information, such as the time shift between two similar signals can be deduced.

In general, it can be shown that when uncorrelated signals are added, the average power of the sum is equal to the sum of the average powers of the signals.

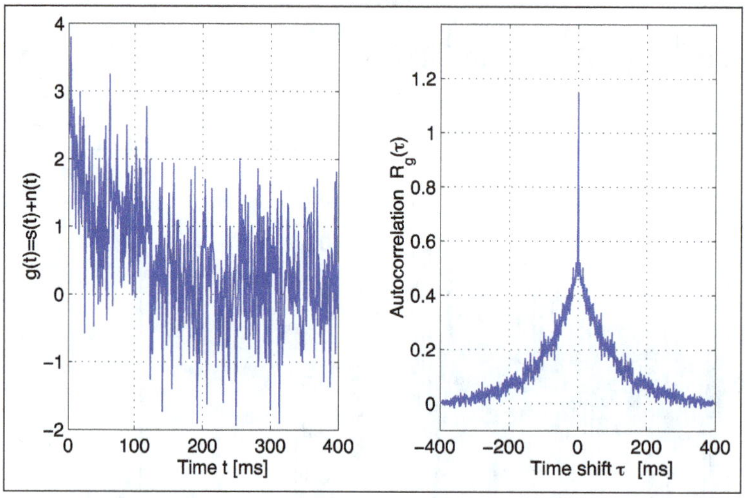

Figure: Time signal and auto-correlation function of exponential waveform plus noise.

A useful application of the correlation function is the detection of signals masked by additive noise. Figure shows the time waveform g(t) which results when the random noise n(t) is added to the signal s(t) of figure. It is seen that the signal waveform is immersed in a considerable amount of noise and it would be difficult to detect s(t) out of g(t). However, the auto-correlation function $R_s(\tau)$ can clearly be recognized from the auto-correlation of the summed signals $R_{s+n}(\tau)$. The detection of signals masked by additive noise is a main issue when dealing with communication systems.

Power Spectral Density

In the following, we will be dealing with truncated signals, i.e. the signals are only considered within a finite time interval [−T/2, T/2] thus assuming the signal to be zero outside this interval. The mathematical representation is not as strait forward as for infinite time signals, however, practical consideration show that this extra effort is needed. Parseval's theorem for truncated signals state that:

$$\int_{-T/2}^{T/2} |f(t)|^2 \, dt = \int_{-\infty}^{\infty} |F_T(f)|^2 \, df$$

Noting the similarity between the first term in previous equation and the time averaged power P of a signal as given in $\overline{n^2(t)} = E\{n^2(t)\} = \lim_{T\to\infty} \frac{1}{T}\int_{-T/2}^{T/2} |n(t)|^2 \, dt$ we write:

$$P = \lim_{T\to\infty} \frac{1}{T}\int_{-T/2}^{T/2} |f(t)|^2 \, dt = \lim_{T\to\infty} \frac{1}{T}\int_{-\infty}^{\infty} |F_T(f)|^2 \, df$$

The first integral in the equation above is easy to understand, it shows that in order to obtain the total power of a signal we must add together the power contribution of each time increment, which is done through the integration over time t. Whether we are dealing with voltage or current is indifferent if we assume a 1 Ω resistance. We know, that evaluating the integral over a finite time period, will give us the signal power within this time. The first integral is thus also valid over each time interval. However, Parseval's equation suggests a second procedure to calculate the total power, which is performed in the frequency domain. The last term in previous equation shows that the summation of $|F(f)|^2$ over all frequencies f will also result in the total power P. Defining a power spectral density function S(f) in units of Watts per Hz such that its integral over frequency is equal to the total power, gives:

$$\int_{-\infty}^{\infty} S(f)df = \lim_{T\to\infty}\frac{1}{T}\int_{-\infty}^{\infty}|F_T(f)|^2 df$$

In addition, we insist that S(f) also gives the power over each frequency increment, which means that the integration of the power density function over a frequency range Δf will give the total power for this frequency interval. It can be shown, that under certain conditions – which are fulfilled for most practical signals of interest – S(f) is related to $|F_T(f)|^2$ through:

$$S(f) = \lim_{T\to\infty}\frac{|F_T(f)|^2}{T}$$

Through $\mathcal{F}\tau\{R_f(\tau)\} = \lim_{T\to\infty}\frac{1}{T}F_T^*(f)F_T(f) = \lim_{T\to\infty}\frac{1}{T}|F_T(f)|^2$ the relation between the power spectral density and the Fourier transform of the auto-correlation function is given:

$$S(f) = \mathcal{F}\{R_f(\tau)\}$$

When evaluating the distribution of noise power over frequency, the power spectral density S(f) should be the function to examine, rather than the Fourier transform. The reason for this, is, that the Fourier transform of a random quantity is also a random quantity, which in this sense does not give us any useful information. As we know, for random signals we need to investigate the statistical properties. Thus in case we are interested in the frequency content of noise, we compute the Fourier transform of the auto-correlation function as given in the expression above.

Using $P = \lim_{T\to\infty}\frac{1}{T}\int_{-T/2}^{T/2}|f(t)|^2 dt = \lim_{T\to\infty}\frac{1}{T}\int_{-\infty}^{\infty}|F_T(f)|^2 df$, $\int_{-\infty}^{\infty} S(f)df = \lim_{T\to\infty}\frac{1}{T}\int_{-\infty}^{\infty}|F_T(f)|^2 df$

and $E\{f^*(t)f(t+\tau)\} = R_f(\tau) = \lim_{T\to\infty}\frac{1}{T}\int_{-1/T}^{1/T} f^*(t)f(t+\tau)dt$ we get:

$$P = \overline{f^2(t)} = \lim_{T\to\infty}\frac{1}{T}\int_{-T/2}^{T/2}|f(t)|^2 dt = \int_{-\infty}^{\infty} S(f)df = R_f(0)$$

The above expression can be used to calculate the total power using the time domain signal, the power spectral density function, or the auto-correlation function. Care must be taken, when the resistive scaling factor is not equal to 1 Ω.

Noise in Linear Systems

When designing and characterizing communication systems, noise is an important parameter which must be accounted for. In general, the different physical noise sources in addition to other man-made noise sources contribute to the total noise in the system. In the following, noise in linear time invariant (LTI) systems is investigated.

Band Limited White Noise

The power spectral density shall be used to describe noise. Knowing that random noise tends to have rapid fluctuations, we assume a noise voltage n (t) having the auto-correlation function:

$$R_n(\tau) = \frac{N_0}{2}\delta(\tau)$$

Where $\delta(\tau)$ is the impulse function. Thus, $R_n(\tau)$ is zero for all $\tau \neq 0$, which indicates completely uncorrelated noise signal except for zero time shift. Taking the Fourier transform of $R_n(\tau)$ the power spectral density is:

$$S_n(f) = \mathcal{F}\{R_n(\tau)\} = N_0/2 \; [\text{Watt/Hz}]$$

The power spectral density is constant for all frequencies, thus it contains all frequency components with equal power weighting. This type of noise is designated as white noise in analogy to white light. The factor of one-half in previous equation is necessary to have a two-sided power spectral density. A problem arises when we try to calculate the total power of white noise, since:

$$P_n = \int_{-\infty}^{\infty} \frac{N_0}{2} df \to \infty$$

Which implies an infinite amount of power and thus cannot be used to describe any physical process. However, it turns out to be a good model for many cases in which the bandwidth is limited through the system. In this case the power spectral density can be assumed flat within the finite measuring bandwidth, which will restrict the total noise power. What we are dealing with in this case is band-limited white noise which will appear as white noise to the measuring system.

The power of band-limited white noise is independent of the choice of operating frequency f_0. If n(t) is zero-mean white noise with the power spectral density equal to $N_0/2$ Watts per Hz, then across a bandwidth B the noise power is:

$$P_n = \int_{f_0-B/2}^{f_0+B/2} \frac{N_0}{2} df + \int_{-f_0-B/2}^{-f_0+B/2} \frac{N_0}{2} df = 2\int_{-B/2}^{B/2} \frac{N_0}{2} df = BN_0 \text{ Watt}$$

Transmission of Noise through an LTI System

The transformation of an input signal x(t) through a linear time invariant (LTI) system is described in the time domain through the convolution integral:

$$y(t) = \int_{-\infty}^{\infty} x(t)h(t-\tau)d\tau$$

Where, y(t) is the output signal and h(t) is the impulse response of the LTI system. If the input signal is random, what we are interested in is the power spectral density $S_y(f)$ of the output signal. Substituting $y(t) = \int_{-\infty}^{\infty} x(t)h(t-\tau)d\tau$ into $E\{f^*(t)f(t+\tau)\} = R_f(\tau) = \lim_{T\to\infty}\frac{1}{T}\int_{-1/T}^{1/T} f^*(t)f(t+\tau)dt$

and performing a transformation of variables we obtain the auto-correlation function of the output signal:

$$R_y(\tau) = \int_{-\infty}^{\infty}\int_{-\infty}^{\infty} R_x(\tau+\alpha-\beta)h*(\alpha)h(\beta)d\alpha d\beta$$

from which $S_y(f)$ is obtained through $S(f)=\mathcal{F}\{R_f(\tau)\}$:

$$S_y(f)=\mathcal{F}\{R_y(\tau)\}=S_x(f)|H(f)|^2$$

Thus, we have the important result that the power density spectrum of the output signal is the product of the input power density spectrum multiplied by the magnitude squared of the frequency transfer function. If the auto-correlation function is desired it is usually easier to compute the power density spectrum through previous equation and then perform the inverse transform:

$$R_y(\tau)=\mathcal{F}^{-1}\{S_y(f)\}=\mathcal{F}^{-1}\left\{S_x(f)|H(f)|^2\right\}$$

If the random input signal is white noise ni(t) with a power spectral density $N_o/2$, then $S_y(f)=\mathcal{F}\{R_y(\tau)\}=S_x(f)|H(f)|^2$ becomes:

$$S_{out}(f) = \frac{N_0}{2}|H(f)|^2$$

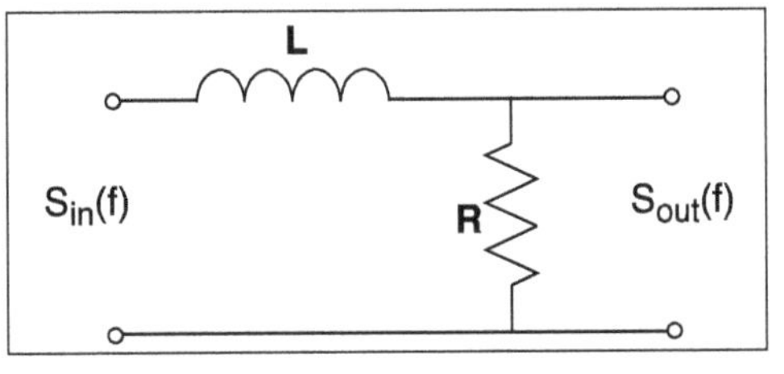

Equivalent Noise Bandwidth

The total noise output power from a system with known frequency transfer function |H(f)| can be calculated using $S_y(f)=\mathcal{F}\{R_y(\tau)\}=S_x(f)|H(f)|^2$

and $P=\overline{f^2(t)}=\lim_{T\to\infty}\frac{1}{T}\int_{-T/2}^{T/2}|f(t)|^2\,dt=\int_{-\infty}^{\infty}S(f)df=R_f(0)$. If the input noise is white, this becomes:

$$P_{out} = \int_{-\infty}^{\infty}S_{out}(f)df = N_o\int_0^{\infty}|H(f)|^2\,df$$

The integral is a constant for a given system frequency transfer function. We would like to have a simple expression similar to $P_n=\int_{f_0-B/2}^{f_0+B/2}\frac{N_0}{2}df+\int_{-f_0-B/2}^{-f_0+B/2}\frac{N_0}{2}df=2\int_{-B/2}^{B/2}\frac{N_0}{2}df=BN_0$ Watt for the output noise power. A reasonable approach would be to define an equivalent noise bandwidth B_N of an ideal filter such that the output noise power from the ideal filter and the real system are equal. We assume that the ideal filters frequency transfer function is flat and equal to $H(f_o)$ within the bandwidth B_N around the center frequency f_o and zero otherwise.

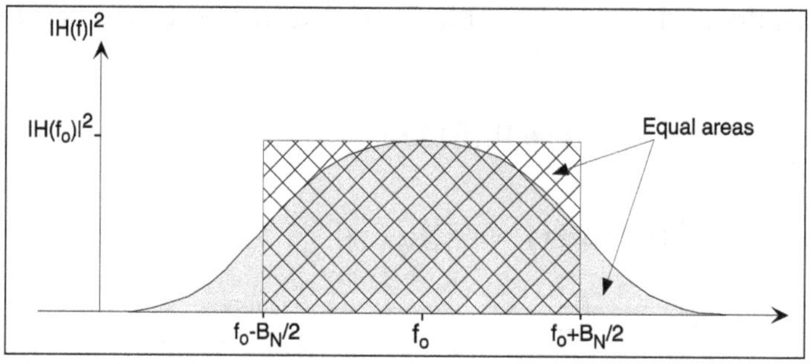

Figure: A graphical definition of the equivalent noise bandwidth.

Thus, the output noise power of the ideal filter is:

$$P_{out} = N_0 \int_{f_0 - B_N/2}^{f_0 + B_N/2} \left|H(f_0)^2\right| df$$

$$= N_0 \left|H(f_0)\right|^2 (f_0 + B_N/2 - f_0 + B_N/2)$$

$$= N_0 \left|H(f_0)\right|^2 B_N$$

Equating the right hand sides of last two equations, we have:

$$B_N = \frac{\int_0^{\infty} \left|H(f)\right|^2 df}{\left|H(f_0)\right|^2} \quad [\text{Hz}]$$

This definition of the equivalent noise bandwidth B_N allows us to discuss practical linear systems by using their idealized equivalents.

Signal-to-Noise Ratio

Let the input signal power for a given device be $\overline{S_{in}^2(t)}$ and let the input noise power of the device be $\overline{n_{in}^2(t)}$. The input signal-to-noise ratio SNR_{in} is defined as the ratio of the total available signal power to the total available noise power at the input and is given by,

$$SNR_{in} = \frac{\overline{s_{in}^2(t)}}{\overline{n_{in}^2(t)}}$$

Thus the SNR as defined above gives an indication of the amount of noise power relative to the signal power. Clearly the signal-to-noise ratio at the output of the device is analog to the above expression. Also the definition of the signal-to-noise ratio is independent of the noise source and type.

The SNR is a power ratio which is most often expressed in Decibels:

$$SNR^{dB} = 10 \log_{10}(SNR)$$

Thus an SNR = 13 dB means that the signal power is twenty times higher than the noise power, while SNR = 0 dB means equal signal and noise power.

Naturally Occurring Noise

Natural radio noise in telecommunication systems is both picked up by the antenna as well as generated within the system itself. The first effect can be accounted for by the contribution which it makes to the antenna noise temperature. Attenuation due to water vapor and oxygen, clouds and precipitation is accompanied by thermal noise, lightening and other atmospherics which further degrades the applicable signal-to-noise ratio. In addition, extraterrestrial noise of thermal or non-thermal origin may be picked up by the receiving antenna.

This section gives an overview of the different types of naturally occurring noise and defines appropriate quantities for modeling the effect of this noise. To start with, we state Planck's radiation law, which is the basis for other types of noise.

Planck's Law

In 1900, Max Planck found the law that governs the emission of electromagnetic radiation from a black body in thermal equilibrium. A black body is simply defined as an idealized, perfectly opaque material that absorbs all the incident radiation at all frequencies, reflecting none. A body in thermodynamic equilibrium emits to its environment the same amount of energy it absorbs from its environment. Hence, in addition to being a perfect absorber, a blackbody also is a perfect emitter. The essential point of Planck's derivation is that energy can only be exchanged in discrete portions or quanta equal to h f, where h is Planck's constant $h = 6.626 \times 10^{-34}$ J s and f is the frequency in Hertz.

Then, the energy of the ground level (or state) is 0, of the first level h f, of the second level 2hf and so on. In general:

$$E_v = n \cdot hf \text{ for } v = 0, 1, 2 \ldots$$

Where, v is the level or state number. Given the number Nv of energy quanta occupying level v results in an energy of $vN_v hf$ for that level. The total energy is obtained by summing up over all states, thus,

$$E_{tot} = N_0 \cdot 0 + N_1 hf + N_2 \cdot 2hf + N_3 \cdot 3hf + \ldots$$

Now according to quantum mechanics, the probability of occupying an energy level goes down with $e^{-\Delta E/kT}$ where $k = 1.38 \times 10^{-23}$ J K^{-1} is the Boltzmann constant, T is the absolute temperature in Kelvin, and ΔE the excess energy. Then, the number of energy quanta N_1 at the first level is given by the number at ground state N_0 multiplied by the probability $e^{-hf/kT}$. Similarly $N_2 = N_1 e^{-hf/kT} = N_0 e^{-2hf/kT}$ and so on. The total number of quanta is:

$$N_{tot} = N_0\left(1 + e^{-hf/kT} + e^{-2hf/kT} + e^{-3hf/kT} + \ldots\right)$$

To determine the average energy, we divide the total energy by the total number of energy quanta. The expression can be simplified to give:

$$\overline{E(f)} = \frac{hf}{e^{hf/kT} - 1}$$

Using the density of modes we find Planck's law for the black body radiation. Expressed in terms of the brightness of the radiated energy from a blackbody this is given by:

$$B_f(f) = \frac{2hf^3}{c^2}\frac{1}{e^{hf/kT}-1}$$

Thermal Radiation

Thermal radiation is system inherent and is generated through the random thermal motion of electrons in a conducting medium such as a resistor. The path of each electron is randomly oriented due to interaction with other electrons. The net effect of the electron motion is a random current flowing in the conduction medium with an average value of zero. The power spectral density of thermal noise is given by Planck's distribution law $\overline{E(f)} = \frac{hf}{e^{hf/kT}-1}$. For the normal range of Temperatures and frequencies well below the optical range the parameter hf/kT is very small, so that $e^{hf/kT} \approx 1 + hf/kT$, and $\overline{E(f)} = \frac{hf}{e^{hf/kT}-1}$ can be approximated by:

$$S_n(f) = kT$$

The power spectral density as given by $S_n(f) = kT$ is independent of frequency and hence is referred to as white noise spectrum. Within the bandwidth B the available noise power then is

$$P_n = kTB$$

The above expression shows that if the bandwidth is fixed it is sufficient to know the temperature in order to be able to compute the noise power. This is the reason, why it is common to speak of the noise temperature when referring to the noise power (even if the noise source is not thermal).

For T =300 K, i.e. at room temperature, we get a noise power of $N_T = -114$ dBm per MHz bandwidth. It is worth remembering this number as a reference and using it to compute the approximate noise power for a given bandwidth. For example the noise power for a 20 MHz system would be $N_T = -101$ dBm.

Knowing the available power to the network, we want to define the circuit equivalent of the noisy resistor. This is done by considering a voltage source of rms voltage V_n connected in series with the resistor R.

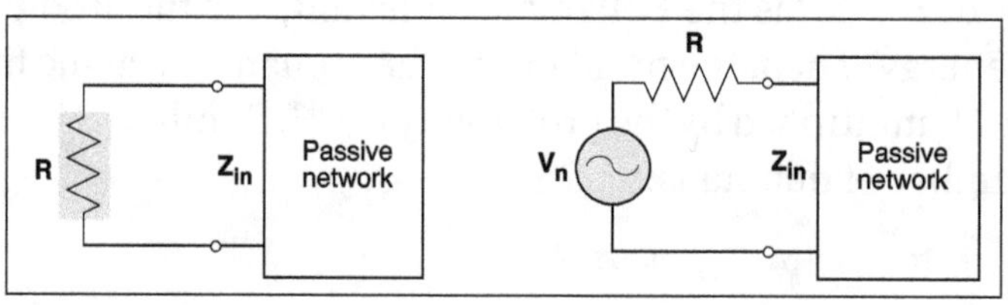

Figure: Noisy resistor connected to a network (left) and its equivalent circuit (right).

The noise power delivered to a network of input impedance Z_{in} is:

$$P_n = \left|\frac{V_n}{R + Z_{in}}\right|^2 R_{in}$$

Where, R_{in} is the resistive component of Z_{in}. If Z_{in} = R, which is the condition for maximum power transfer, we find that,

$$P_n = \frac{V_n^2}{4R}$$

Substituting for P_n from $P_n = kTB$ and solving for V_n gives,

$$V_n = \sqrt{\overline{v_n^2(t)}} = 2\sqrt{kTRB}$$

Extraterrestrial Noise

Space is the source of mostly broadband noise which can be considered as plane electromagnetic waves. Cosmic radiation has to be accounted for if either the main lobe or the side lobes of the receiving antenna are directed towards space. The noise sources are both thermal and non-thermal emission from the Sun, the Moon, the Cassiopeia and planets and from elsewhere in our galaxy and other galaxies.

If the emission is of thermal origin its contribution to noise power can be described through the spectral brightness as given in $B_f(f) = \frac{2hf^3}{c^2} \frac{1}{e^{hf/kT} - 1}$ which is the power density in Watt per unit solid angle per unit area per Hertz. At radio-frequencies where hf ≪ kT the spectral brightness Bf is given by the Rayleigh-Jeans law:

$$B_f = \frac{2kT_c}{\lambda^2} \quad \text{in} \quad \left[\frac{\text{Watt}}{m^2 \cdot sr \cdot \text{Hertz}}\right]$$

Where,

- T_c is the brightness temperature.
- λ is the wavelength.
- k is Boltzmann's constant.

The actual noise power received within a narrow frequency range depends on the direction of the main lobe and the side lobes of the receiving antenna and on the effective area of the antenna. Thus in general the spectral brightness of an extended source is a function of the direction relative to the antenna coordinates. For discrete sources (such as the Sun), which lie within the main lobe of the antenna and subtend a solid angle Ω_s that is much smaller than the antenna main-beam solid angle, the spectral power density becomes,

$$p = \frac{2kT_c}{\lambda^2}\Omega_s \quad Wm^{-2}Hz$$

In the general case B_f varies as λ^n where n is known as the spectral index. Thus for the thermal emission of a black body n = −2. For non-thermal emission $B_f = \frac{2kT_c}{\lambda^2}$ in $\left[\frac{\text{Watt}}{m^2 \cdot sr \cdot \text{Hertz}}\right]$ can still be used but the brightness temperature T_c is no longer related to the thermal emission but is an equivalent brightness temperature, in addition the spectral index has to be specified.

Background Radiation

The entire Universe is saturated with what is known as microwave background radiation, a remnant of the Big Bang. After the Big Bang, the formation of matter, space and time out of virtually nothing, the prevailing temperatures were at first almost inconceivably high. However, as the Universe expanded the temperature sank to approximately −270 °C, the temperature that it is today. The expansion of space lengthened the wavelength of the electromagnetic radiation until it entered the microwave range. Today, this radiation can be measured reaching us evenly from all directions of space, thus the term "background radiation". It would "heat up" any colder object to the space temperature of 3 K (note that absolute zero Kelvin is −273 °C).

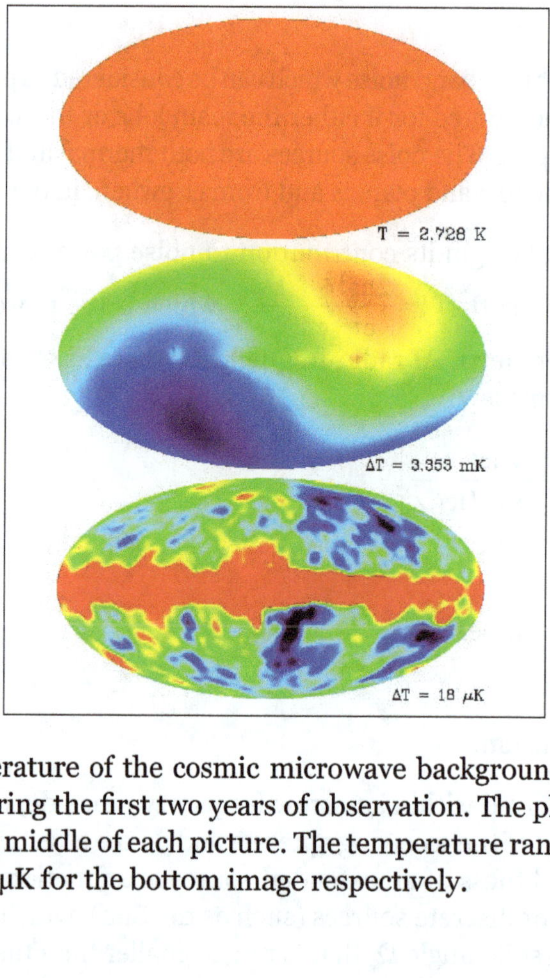

As shown in figure Temperature of the cosmic microwave background radiation as determined with the COBE satellite during the first two years of observation. The plane of the Milky Way Galaxy is horizontal across the middle of each picture. The temperature range is 0 – 4K for the top, 3.3 mK for the middle, and 18 μK for the bottom image respectively.

Absorption Noise

When energy is absorbed by a body the same energy is reradiated as noise as shown by the theory of black body emission. Otherwise the temperature of some bodies would rise and that of others fall. In the case of a radiating antenna the energy is partially absorbed by the atmosphere and reradiated as noise. The effective absorption noise temperature T_{ab} given as a function of the ambient temperature T_a and the attenuation L_a is:

$$T_{ab} = T_a (L_a - 1)$$

T_{ab} is not identical to the physical (ambient) temperature of the atmosphere and increases with increasing atmospheric attenuation. Table shows some values for L_a and T_{ab} when the ambient temperature T_a is 300K.

Table: Absorption noise temperature T_{ab} for different values of attenuation and an ambient temperature T_a = 300K

L_a [dB]	0	1	3	10
L_a (power ratio)	1	1.26	2	10
T_{ab} [K]	0	78	300	2700

The Attenuation of the atmosphere depends strongly on frequency. Water vapor and oxygen cause a high atmospheric attenuation in the 20GHz and 60GHz frequency bands. The frequency band below 10GHz exhibits the lowest noise temperature.

Frequency-response Model

A frequency-response model is the frequency response of a linear system evaluated over a range of frequency values. The model is represented by an idfrd model object that stores the frequency response, sample time, and input-output channel information.

The frequency-response function describes the steady-state response of a system to sinusoidal inputs. For a linear system, a sinusoidal input of a specific frequency results in an output that is also a sinusoid with the same frequency, but with a different amplitude and phase. The frequency-response function describes the amplitude change and phase shift as a function of frequency. You can estimate frequency-response models and visualize the responses on a Bode plot, which shows the amplitude change and the phase shift as a function of the sinusoid frequency.

For a discrete-time system sampled with a time interval T, the transfer function G(z) relates the Z-transforms of the input U(z) and output Y(z):

$$Y(z)=G(z)U(z)+H(z)E(z)$$

The frequency-response is the value of the transfer function, G(z), evaluated on the unit circle (z = $\exp^{iwT}$) for a vector of frequencies, w. H(z) represents the noise transfer function, and E(z) is the Z-transform of the additive disturbance e(t) with variance λ. The values of G are stored in the Response Data property of the idfrd object. The noise spectrum is stored in the SpectrumData property.

Where, the noise spectrum is defined as:

$$\Phi_v(\omega)=\lambda T\left|H\left(e^{i\omega T}\right)\right|^2$$

A MIMO frequency-response model contains frequency-responses corresponding to each input-output pair in the system. For example, for a two-input, two-output model:

$$Y_1(z)=G_{11}(z)U_1(z)+G_{12}(z)U_2(z)+H_1(z)E_1(z)$$
$$Y_2(z)=G_{21}(z)U_1(z)+G_{22}(z)U_2(z)+H_2(z)E_2(z)$$

Where, G_{ij} is the transfer function between the ith output and the jth input. $H_1(z)$ and $H_2(z)$ represent the noise transfer functions for the two outputs. $E_1(z)$ and $E_2(z)$ are the Z-transforms of the additive disturbances, $e_1(t)$ and $e_2(t)$, at the two model outputs, respectively.

Similar expressions apply for continuous-time frequency response. The equations are represented in Laplace domain.

References

- What-is-multiplexing-types-and-their-applications: elprocus.com, Retrieved 23, June 2020
- Communication-channels: mangementchallenge.weebly.com, Retrieved 05, April 2020
- Multiplexing-in-computer-network: javatpoint.com, Retrieved 04, February 2020
- Return-channel, computing: encyclopedia.com, Retrieved 20, July 2020
- What-is-audio-return-channel-arc-and-how-it-works: aovers.com, Retrieved 11, March 2020
- Difference-between-upstream-and-downstream-on-a-cable-modem: techwalla.com, Retrieved 05, May 2020

3

Understanding Wireless Network

The computer networks which are not connected by cables or wires are called wireless networks. There are various types of wireless networks such as wireless personal area network (PAN), wireless local area network (LAN), wireless ad hoc network, wireless metropolitian network (MAN), wireless wide area network (WAN), cellular network and global area network, etc. This chapter has been carefully written to provide an easy understanding of the varied facets of a wireless network.

A wireless network allows devices to stay connected to the network but roam untethered to any wires. Access points amplify Wi-Fi signals, so a device can be far from a router but still be connected to the network. When you connect to a Wi-Fi hotspot at a cafe, a hotel, an airport lounge, or another public place, you're connecting to that business's wireless network. A wired network uses cables to connect devices, such as laptop or desktop computers, to the Internet or another network. A wired network has some disadvantages when compared to a wireless network. The biggest disadvantage is that your device is tethered to a router. The most common wired networks use cables connected at one end to an Ethernet port on the network router and at the other end to a computer or other device.

How Wireless Networks Work

Moving data through a wireless network involves three separate elements: the radio signals, the data format, and the network structure. Each of these elements is independent of the other two, so you must define all three when you invent a new network. In terms of the OSI reference model, the radio signal operates at the physical layer, and the data format controls several of the higher layers. The network structure includes the wireless network interface adapters and base stations that send and receive the radio signals. In a wireless network, the network interface adapters in each computer and base station convert digital data to radio signals, which they transmit to other devices on the same network, and they receive and convert incoming radio signals from other network elements back to digital data. Each of the broadband wireless data services use a different combination of radio signals, data formats, and network structure.

Radio

The basic physical laws that make radio possible are known as Maxwell's equations, identified by James Clerk Maxwell in 1864. Without going into the math, Maxwell's equations show that a changing magnetic field will produce an electric field, and a changing electric field will produce a magnetic field. When alternating current (AC) moves through a wire or other physical conductor, some of that energy escapes into the surrounding space as an alternating magnetic field. That magnetic field creates an alternating electric field in space, which in turn creates another magnetic field and so forth until the original current is interrupted.

This form of energy in transition between electricity and magnetic energy is called electromagnetic radiation, or radio waves. Radio is defined as the radiation of electromagnetic energy through space. A device that produces radio waves is called a transmitter, and a complementary device that detects radio waves in the air and converts them to some other form of energy is called a receiver. Both transmitters and receivers use specially shaped devices called antennas to focus the radio signal in a particular direction, or pattern, and to increase the amount of effective radiation (from a transmitter) or sensitivity (in a receiver).

By adjusting the rate at which alternating current flows from each transmitter through the antenna and out into space (the frequency), and by adjusting a receiver to operate only at that frequency, it's possible to send and receive many different signals, each at a different frequency, that don't interfere with one another. The overall range of frequencies is known as the radio spectrum. A smaller segment of the radio spectrum is often called a band.

Radio frequencies and other AC signals are expressed as cycles per second, or hertz (Hz), named for Heinrich Hertz, the first experimenter to send and receive radio waves. One cycle is the distance from the peak of an AC signal to the peak of the next signal. Radio signals generally operate at frequencies in thousands, millions, or billions of hertz (kilohertz or KHz, megahertz or MHz, and gigahertz or GHz, respectively).

The simplest type of radio communication uses a continuous signal that the operator of the transmitter interrupts to divide the signal into accepted patterns of long and short signals (dots and dashes) that correspond to individual letters and other characters. The most widely used set of these patterns was Morse code, named for the inventor of the telegraph, Samuel F.B. Morse, where this code was first used.

In order to transmit speech, music, and other sounds via radio, the transmitter alters, or modulates, the AC signal (the carrier wave) by either mixing an audio signal with the carrier as shown in figure (this is called amplitude modulation, or AM) or by modulating the frequency within a narrow range as shown in Figure below (this is called frequency modulation, or FM). The AM or FM receiver includes a complementary circuit that separates the carrier from the modulating signal.

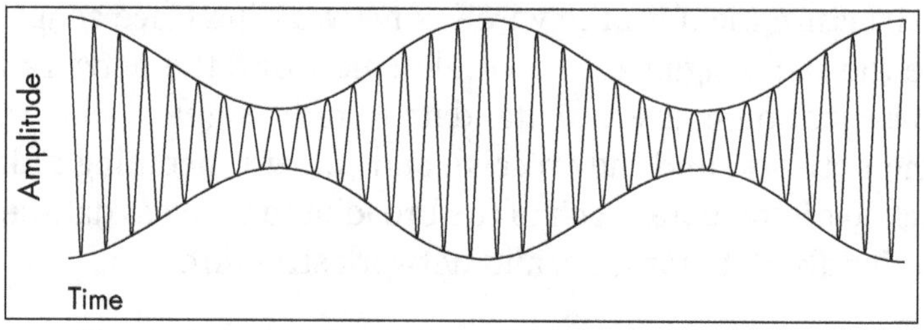

Figure: In an AM signal, the audio modulates the carrier.

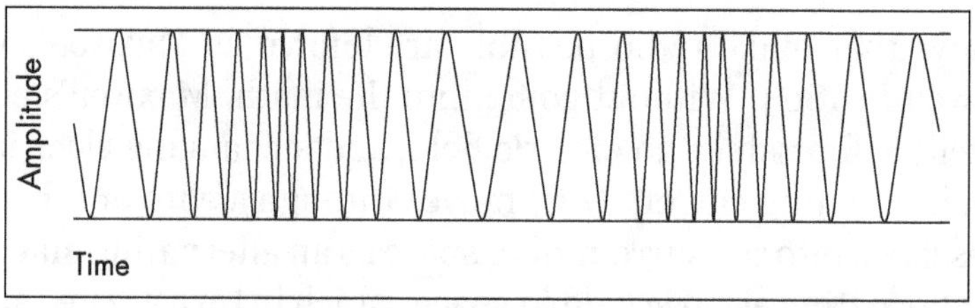

Figure: In an FM signal, the audio modulates the radio frequency.

Because two or more radio signals using the same frequency can often interfere with one another, government regulators and international agencies, such as the International Telecommunication

Union (ITU), have reserved certain frequencies for specific types of modulation, and they issue exclusive licenses to individual users. For example, an FM radio station might be licensed to operate at 92.1 MHz at a certain geographical location. Nobody else is allowed to use that frequency in close enough proximity to interfere with that signal. On the other hand, some radio services don't require a license. Most unlicensed services are either restricted to very short distances, to specific frequency bands, or both.

Both AM and FM are analog methods because the signal that comes out of the receiver is a replica of the signal that went into the transmitter. When we send computer data through a radio link, it's digital because the content has been converted from text, computer code, sounds, images or other information into ones and zeroes before it is transmitted, and it is converted back to its original form after it is received. Digital radio can use any of several different modulation methods: The ones and zeroes can be two different audio tones, two different radio frequencies, timed interruptions to the carrier, or some combination of those and other techniques.

Wireless Data Networks

Each type of wireless data network operates on a specific set of radio frequencies. For example, most Wi-Fi networks operate in a special band of radio frequencies around 2.4 GHz that have been reserved in most parts of the world for unlicensed point-to-point spread spectrum radio services. Other Wi-Fi systems use a different unlicensed band around 5 GHz.

Unlicensed Radio Services

Unlicensed means that anybody using equipment that complies with the technical requirements can send and receive radio signals on these frequencies without a radio station license. Unlike most radio services (including other broadband wireless services), which require licenses that grant exclusive use of that frequency to a specific type of service and to one or more specific users, an unlicensed service is a free-for-all where everybody has an equal claim to the same airwaves. In theory, the technology of spread spectrum radio makes it possible for many users to co-exist (up to a point) without significant interference.

Point-to-Point

A point-to-point radio service operates a communication channel that carries information from a transmitter to a single receiver. The opposite of point-to point is a broadcast service (such as a radio or television station) that sends the same signal to many receivers at the same time.

Spread Spectrum

Spread spectrum is a family of methods for transmitting a single radio signal using a relatively wide segment of the radio spectrum. Wireless Ethernet networks use several different spread spectrum radio transmission systems, which are called frequency-hopping spread spectrum (FHSS), direct-sequence spread spectrum (DSSS), and orthogonal frequency division multiplexing (OFDM). Some older data networks use the slower FHSS system, but the first Wi-Fi networks used DSSS, and more recent systems use OFDM. Table lists each of the Wi-Fi standards and the type of spread spectrum modulation they use.

Table: Wi-Fi Standards and Modulation Type.

Wi-Fi Type	Frequency	Modulation
802.11a	5 GHz	OFDM
802.11b	2.4 GHz	DSSS
802.11g	2.4 GHz	OFDM

Spread spectrum radio offers some important advantages over other types of radio signals that use a single narrow channel. Spread spectrum is extremely efficient, so the radio transmitters can operate with very low power. Because the signals operate on a relatively wide band of frequencies, they are less sensitive to interference from other radio signals and electrical noise, which means they can often get through in environments where a conventional narrow-band signal would be impossible to receive and understand. And because a frequency-hopping spread spectrum signal shifts among more than one channel, it can be extremely difficult for an unauthorized listener to intercept and decode the contents of a signal.

Spread spectrum technology has an interesting history. It was invented by the actress Hedy Lamarr and the American avant-garde composer George Antheil as a "Secret Communication System" for directing radio-controlled torpedoes that would not be vulnerable to enemy jamming. Before she came to Hollywood, Lamarr had been married to an arms merchant in Austria, where she learned about the problems of torpedo guidance at dinner parties with her husband's customers. Years later, shortly before the United States entered World War II, she came up with the concept of changing radio frequencies to cut through interference.

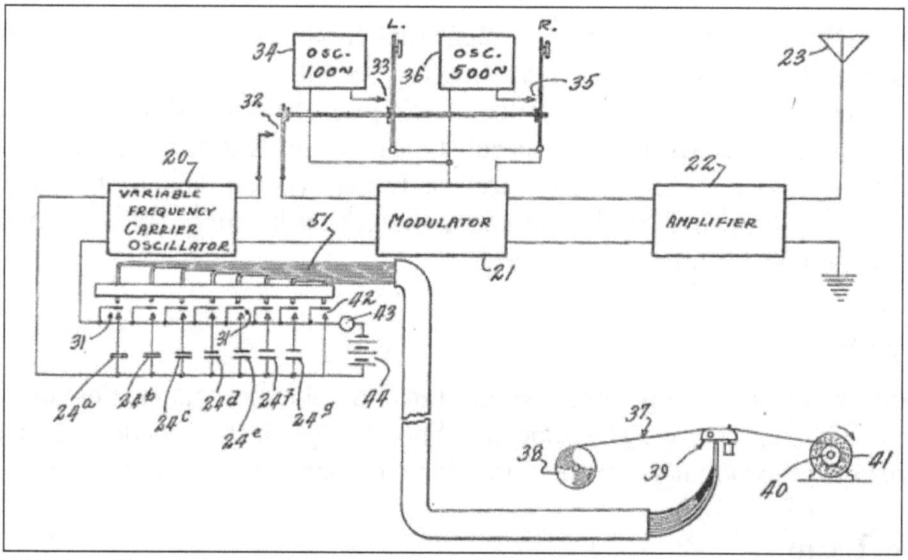

Figure: Hedy Lamarr and George Antheil received this patent in 1942 for the invention that became the foundation of spread spectrum radio communication.

Antheil turned out to be the ideal person to make this idea work. His most famous composition was an extravaganza called Ballet Mechanique, which was scored for sixteen player pianos, two airplane propellers, four xylophones, four bass drums, and a siren. His design used the same kind of mechanism that he had previously used to synchronize the player pianos to change radio frequencies in a spread spectrum transmission. The original slotted paper tape system had 88 different radio channels—one for each of the 88 keys on a piano.

In theory, the same method could be used for voice and data communication as well as guiding torpedoes, but in the days of vacuum tubes, paper tape, and mechanical synchronization, the whole process was too complicated to actually build and use. By 1962, solid-state electronics had replaced the vacuum tubes and piano rolls, and the technology was used aboard US Navy ships for secure communication during the Cuban Missile Crisis. Today, spread spectrum radios are used in the US Air Force Space Command's Milstar Satellite Communications System, in digital cellular telephones, and in wireless data networks.

Frequency-hopping Spread Spectrum

Lamarr and Antheil's original design for spread spectrum radio used a frequency-hopping system (FHSS). As the name suggests, FHSS technology divides a radio signal into small segments and "hops" from one frequency to another many times per second as it transmits those segments. The transmitter and the receiver establish a synchronized hopping pattern that sets the sequence in which they will use different sub channels.

FHSS systems overcome interference from other users by using a narrow carrier signal that changes frequency many times per second. Additional transmitter and receiver pairs can use different hopping patterns on the same set of sub channels at the same time. At any point in time, each transmission is probably using a different sub channel, so there's no interference between signals. When a conflict does occur, the system resends the same packet until the receiver gets a clean copy and sends a confirmation back to the transmitting station.

For some older 802.11 wireless data services, the unlicensed 2.4 MHz band is split into 75 sub channels, each of them 1 MHz wide. Because each frequency hop adds overhead to the data stream, FHSS transmissions are relatively slow.

Direct-sequence Spread Spectrum

The direct-sequence spread spectrum (DSSS) technology that controls 802.11b networks uses an 11-chip Barker Sequence to spread the radio signal through a single 22 MHz–wide channel without changing frequencies. Each DSSS link uses just one channel without any hopping between frequencies. As figure shows, a DSSS transmission uses more bandwidth, but less power than a conventional signal. The digital signal on the left is a conventional transmission in which the power is concentrated within a tight bandwidth. The DSSS signal on the right uses the same amount of power, but it spreads that power across a wider band of radio frequencies. Obviously, the 22 MHz DSSS channel is a lot wider than the 1 MHz channels used in FHSS systems.

A DSSS transmitter breaks each bit in the original data stream into a series of redundant bit patterns called chips, and it transmits them to a receiver that reassembles the chips back into a data stream that is identical to the original. Because most interference is likely to occupy a narrower bandwidth than a DSSS signal, and because each bit is divided into several chips, the receiver can usually identify noise and reject it before it decodes the signal.

Like other networking protocols, a DSSS wireless link exchanges handshaking messages within each data packet to confirm that the receiver can understand each packet. For example, the standard data transmission rate in an 802.11b DSSS Wi-Fi network is 11Mbps, but when the signal quality won't support that speed, the transmitter and receiver use a process called dynamic rate shifting to drop the speed down to 5.5Mbps. The speed might drop because a source of electrical noise near the receiver interferes with the signal or because the transmitter and receiver are too far apart to support full-speed operation. If 5.5Mbps is still too fast for the link to handle, it drops again, down to 2Mbps or even 1Mbps.

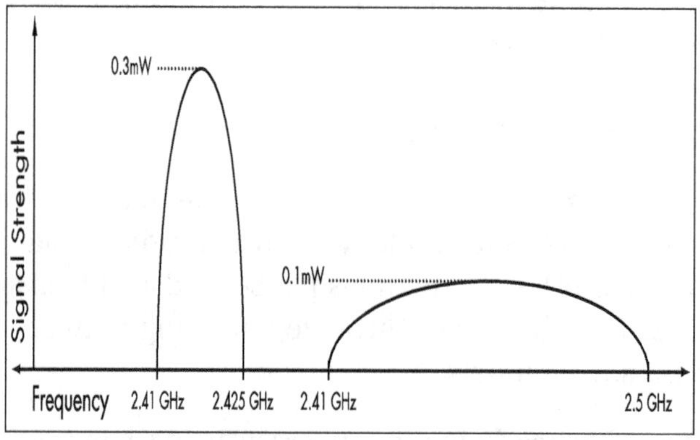

Figure: A conventional signal (left) uses a narrow radio frequency bandwidth. A DSSS signal (right) uses a wider bandwidth but a less powerful signal.

Orthogonal Frequency Division Multiplexing

Orthogonal frequency division multiplexing (OFDM) modulation, used in 802.11a Wi-Fi networks, is considerably more complicated than DSSS technology. The physical layer splits the data stream among 52 parallel bit streams that each use a different radio frequency called a subcarrier. Four of these subcarriers carry pilot data that provides reference information about the remaining 48 subcarriers, in order to reduce signal loss due to radio interference or phase shift. Because the data is divided into 48 separate streams that move through separate subcarriers in parallel, the total transmission speed is much greater than the speed of data through a single channel.

The subcarrier frequencies in an OFDM signal overlap with the peak of each subcarrier's waveform matching the baseline of the overlapping signals as shown in figure. This is called orthogonal frequency division. The 802.11a standard specifies a total of eight data channels that are 20 MHz wide. Each of these channels is divided into 52 300 kHz subcarriers.

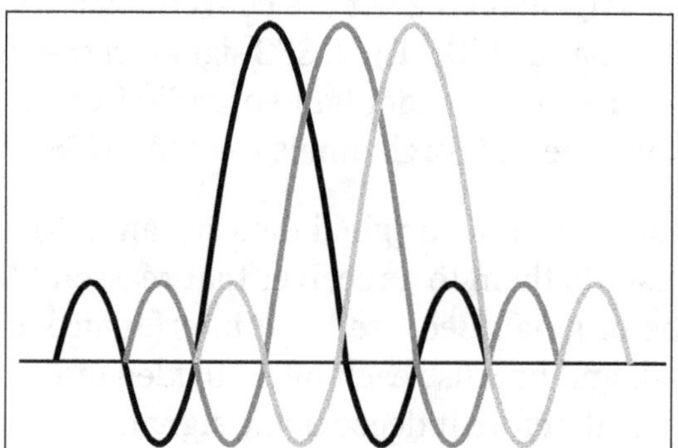

Figure: In OFDM, the peaks of overlapping frequencies don't interfere with one another.

When a Wi-Fi radio receiver detects an 802.11a signal, it assembles the parallel bit streams back into a single high-speed data stream and uses the pilot data to check its accuracy. Under ideal conditions, an 802.11a network can move data at 54Mbps, but like DSSS modulation, the OFDM transmitter and receiver automatically reduces the data speed when the maximum transmission rate is not possible due to interference, weak signals, or other less than-perfect atmospheric conditions.

The more recent 802.11g specification was designed to combine the best features of both 802.11b (greater signal range) and 802.11a (higher speed). To accomplish this objective, it uses OFDM modulation on the 2.4 GHz frequency band.

Why This Matters

The great science fiction writer Arthur C. Clarke once observed that "Any sufficiently advanced technology is indistinguishable from magic." For most of us, the technology that controls high-speed spread spectrum radio could just as easily be a form of magic, because we don't need to understand the things that happen inside a transmitter and a receiver; they're just about invisible when we connect a computer to the Internet.

But when you know that there's a well-defined set of rules and methods that make the connection work (even if you don't know all the details), you are in control. You know that it's not magic, and if you think about it, you might also know some of the right questions to ask when the system doesn't work correctly. If knowledge is power, then knowledge about the technology you use every day is the power to control that technology rather than just use it.

Benefits of Wireless

Wireless broadband provides Internet access to mobile devices in addition to allowing network operators to extend their networks beyond the range of their wired connections. For our purposes, two-way radio is the most sensible approach to wireless broadband, but other methods (such as infrared light or visible signaling) are also possible. Connecting your computer to the Internet (or a local network) by radio offers several advantages over connecting the same computer through a wired connection. First, wireless provides convenient access for portable computers; it's not necessary to find a cable or network data outlet. And second, it allows a user to make a connection from more than one location and to maintain a connection as the user moves from place to place. For network managers, a wireless connection makes it possible to distribute access to a network without the need to string wires or cut holes through walls.

In practice, access without cables means that the owner of a laptop or other portable computer can walk into a classroom, a coffee shop, or a library and connect to the Internet by simply turning on the computer and running a communication program. Depending on the type of wireless network you're using, you might also be able to maintain the same connection in a moving vehicle.

When you're installing your own network, it's often easier to use Wi-Fi links to extend your network and your Internet connection to other rooms because a wired system requires a physical path for the cables between the network router or switch and each computer. Unless you can route those cables through a false ceiling or some other existing channel, this almost always means that you must cut holes in your walls for data connectors and feed wires inside the walls and under the floors. A radio signal that passes through those same walls is often a lot neater and easier.

Wireless Data Services

Because radio signals move through the air, you can set up a network connection from any place within range of the network base station's transmitter; it's not necessary to use a telephone line, television cable, or some other dedicated wiring to connect your computer to the network. Just turn on the radio connected to the computer and it will find the network signal. Therefore, a radio (or wireless) network connection is often a lot more convenient than a wired one.

This is not to say that wireless is always the best choice. A wired network is usually more secure than a wireless system because it's a lot more difficult for unauthorized eavesdroppers and other snoops to monitor data as it moves through the network, and a wired link doesn't require as many complex negotiations between the sender and receiver on protocols and so forth. In an environment where your computer never moves away from your desk and there are no physical obstacles between the computer and the network access point, it's often easier to install a data cable between the computer and a modem.

So now we have a bunch of radio transmitters and receivers that all operate on the same frequencies and all use the same kind of modulation. (Modulation is the method a radio uses to add some kind of content, such as voice or digital data, to a radio wave.) The next step is to send some network data through those radios. Several different wireless data systems and services are available to connect computers and other devices to local networks and to the Internet, including Wi-Fi, WiMAX, and a handful of services based on the latest generations of cellular mobile telephone technology.

Wi-Fi

The IEEE (Institute of Electrical and Electronics Engineers) has produced a set of standards and specifications for wireless networks under the title IEEE 802.11 that define the formats and structures of the relatively short-range signals that provide Wi-Fi service. The original 802.11 standard (without any letter at the end) was released in 1997. It covers several types of wireless media: two kinds of radio transmissions and networks that use infrared light. The 802.11b standard provides additional specifications for wireless Ethernet networks. A related document, IEEE 802.11a, describes wireless networks that operate at higher speeds on different radio frequencies. Still other 802.11 radio networking standards with other letters are also available or moving toward public release.

The specifications in widest use today are 802.11a, 802.11b, and 802.11g. They're the de facto standards used by just about every wireless Ethernet LAN that you are likely to encounter in offices and public spaces and in most home networks. It's worth the trouble to keep an eye on the progress of those other standards, but for the moment, 802.11a and 802.11g are the ones to use for short-range wireless networks, especially if you're expecting to connect to networks where you don't control all the hardware yourself.

The 802.11n standard is the next one in the pipeline, and when it's released, it will replace both 802.11b and 802.11g because it's faster, more secure, and more reliable. The older standards will still work, so new Wi-Fi equipment will support all three (often along with 802.11a, which uses different radio frequencies) and automatically match your network interface to the signals it detects from each base station.

There are two more names in the alphabet soup of wireless LAN standards that you ought to know about: WECA and Wi-Fi. WECA (Wireless Ethernet Compatibility Alliance) is an industry group that includes all of the major manufacturers of wireless Ethernet equipment. Their twin missions are to test and certify that the wireless network devices from all of their member companies can operate together in the same network, and to promote 802.11 networks as the worldwide standard for wireless LANs. WECA's marketing geniuses have adopted the friendlier name of Wi-Fi (short for wireless fidelity) for the 802.11 specifications.

Once or twice per year, the Wi-Fi Alliance conducts an "interoperability bake-off" where engineers from many hardware manufacturers confirm that their hardware will communicate correctly with equipment from other suppliers. Network equipment that carries a Wi-Fi logo has been certified by the Wi-Fi Alliance to meet the relevant standards and to pass interoperability tests. Figure shows one version of the Wi-Fi logo.

Wi-Fi was originally intended to be a wireless extension of a wired LAN, so the distances between Wi-Fi base stations and the computers that communicate through them are limited to about 100 feet (35 meters) indoors or up to 300 feet (100 meters) outdoors, assuming there are no obstructions between the access point and the computer. When 802.11n equipment becomes available, it will support connections between computers and base stations at least as far apart as the older Wi-Fi versions. There are ways to extend the range of a Wi-Fi signal, but those techniques require special equipment and careful installation.

Figure: A Wi-Fi logo.

Because most Wi-Fi signals have such a limited range, you must find a new access point, or hot spot, and set up a new connection every time you move your computer to a new location. And because many Wi-Fi access points don't permit strangers to connect through them, you may have to establish a separate account for each location.

Metropolitan Wi-Fi Services

In some metropolitan areas, a large number of interconnected Wi-Fi base stations are being installed by either local government agencies or private businesses to provide wireless service throughout an entire region or in selected neighborhoods as an economical alternative to cable and telephone (DSL) services. The base stations for these services are often mounted on utility poles or rooftops.

These same networks might also provide a variety of special data services to the local government and major subscribers. For example, the local natural gas, electric, and water utilities could add small Wi-Fi adapters to their meters and use the system to send readings once a month. And city buses might have transponders that report their locations to a central tracking system.

It's not yet clear whether these city-wide Wi-Fi services will be able to overcome possible interference problems and competition from other wireless data alternatives, or whether they will attract enough business to remain viable. But if they do, any computer within the coverage area that has a Wi-Fi adapter should detect the signal and have access to a broadband Internet connection.

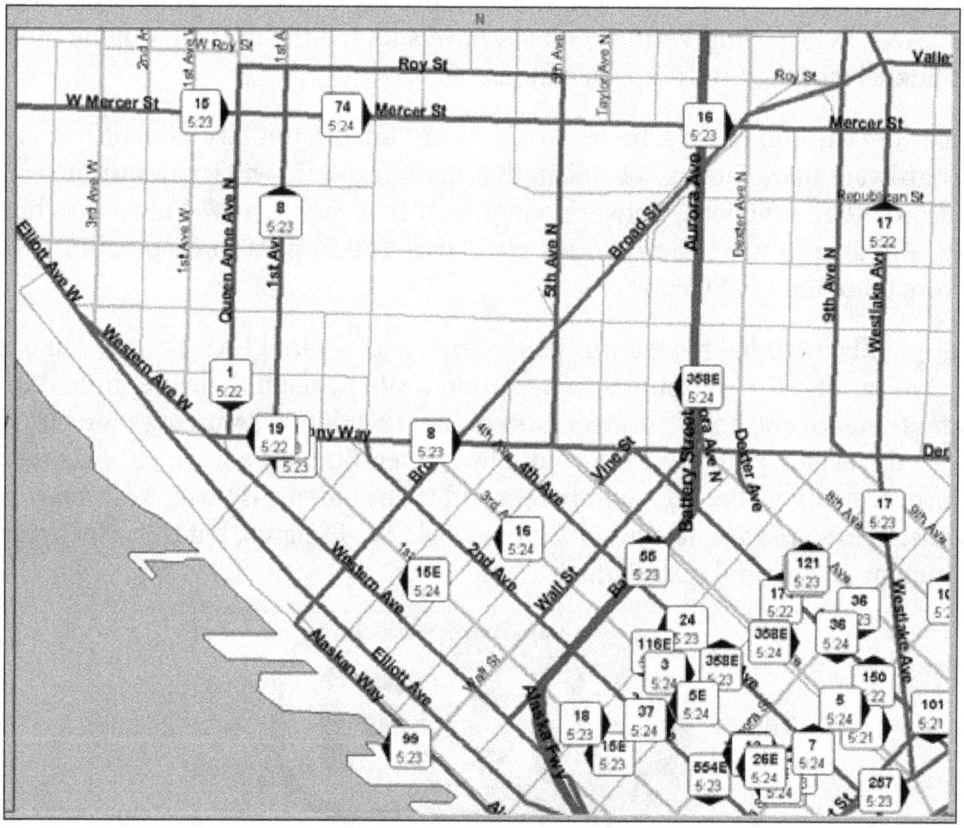

Figure: Wireless technology tracks city buses in Seattle and reports locations on a website.

Cellular Mobile Wireless Services

Several broadband wireless data services are extensions of cellular mobile telephone technology. You might see them described as 3G services because they're based on the third generation of cellular telephone technology. If you have been using a mobile telephone for more than a year or two, you probably remember that the earliest phones were only good for voice calls, but as each new generation was introduced, your mobile carrier offered more and better features. Table describes the various generations.

For people who use their computers away from their home or office, the great advantage of a mobile broadband service is that it covers a much wider territory than any Wi-Fi base station; you can connect your computer to the Internet without the need to search for a new hot spot and use a different access account in each new location, and you can even keep the same connection alive in a moving vehicle. Each of the major wireless broadband services offers coverage in most metropolitan areas and much of the countryside between cities.

Table: Cellular mobile telephone generations.

Name	Features
1G	Analog voice communication only.
2G	System can handle more calls. Digital voice: • Uses less power. • Less background noise. Digital data: • Simple text messages. • Email.
2.5G	Packet-switched signaling. Faster data transfer (up to 144Kbps). Supports relatively slow Internet connections.
3G	Even more calls at the same time. Much faster data transfer rates (up to 2.4Mbps). Broadband Internet. Video and music.
4G	Based on Internet technology. Packet signaling. Very high speed (100Mbps–1Gbps). Will combine telephone, computer, and other technologies.

Of course, computer technology has also been improving at the same time, so today's 2.5G and 3G mobile telephones often incorporate enough computing power to allow them to double as pocket-size Internet terminals (as well as cameras and media players). Broadband data adapters that use 2.5G and 3G technology can attach to a laptop or other portable computer and provide a direct wireless connection to the Internet through the same cellular telephone company that offers mobile telephone service.

Today, most cellular broadband wireless services offer credit card–size adapters that connect to your computer through the PC Card socket on the side of a laptop or into the front or back panel of a desktop computer. In another year or two, many new laptops will come with internal adapters and integrated antennas for both Wi-Fi and 3G wireless or WiMAX that mount directly on the motherboard, just as they contain internal Wi-Fi adapters and dial-up modems today.

WiMAX

Worldwide Interoperability for Microwave Access (WiMAX) is yet another method for distributing broadband wireless data over wide geographic areas. It's a metropolitan area network service that typically uses one or more base stations that can each provide service to users within a 30-mile radius. The IEEE 802.16 specification contains the technical details of WiMAX networks.

In the United States, the earliest WiMAX services were offered by Clearwire as a wireless alternative to DSL and cable broadband Internet access in fixed locations (such as homes and businesses), but mobile WiMAX access is not far behind. By early 2008, Clearwire plans to offer access to their wireless networks through an adapter on a PC Card. When those adapters become available, WiMAX, 3G cellular data services, and metropolitan Wi-Fi networks will compete for the same commercial niche: wireless access to the Internet through a service that covers an entire metropolitan area.

Each WiMAX service provider uses one or more licensed operating frequencies somewhere between 2 GHz and 11 GHz. A WiMAX link can transfer data (including handshaking and other overhead) at up to 70Mbps, but most commercial WiMAX services are significantly slower than that. And as more and more users share a single WiMAX tower and base station, some users report that their signal quality deteriorates.

Unlike the cellular broadband wireless data services that piggyback on existing mobile telephone networks, WiMAX is a separate radio system that is designed to either supplement or replace the existing broadband Internet distribution systems. In practice, WiMAX competes with both 3G wireless services and with Internet service providers that distribute Internet access to fixed locations through telephone lines and cable television utilities. Home and business subscribers to a WiMAX service usually use either a wired LAN or Wi-Fi to distribute the network within their buildings. Figure below shows a typical WiMAX network.

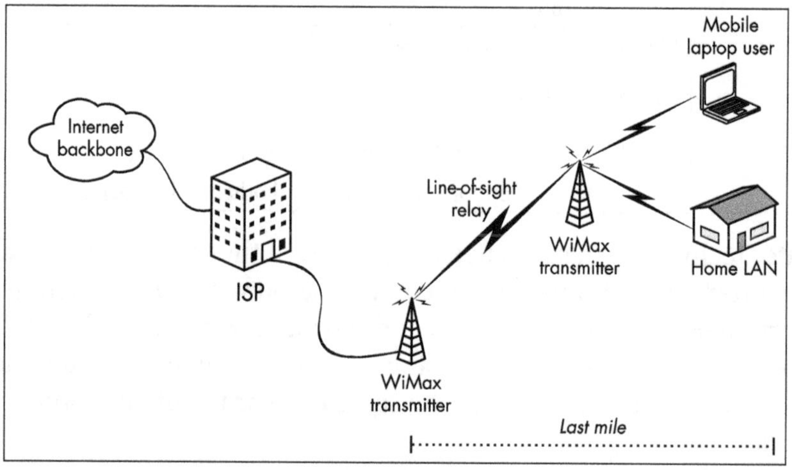

Figure: WiMAX provides last mile Internet connections to homes and businesses.

Bluetooth

Bluetooth is the other type of wireless networking technology that we ought to describe. Bluetooth uses radio signals to replace the wires and cables that connect a computer or a mobile telephone to peripheral devices, such as a keyboard, a mouse, or a set of speakers. You can also use Bluetooth to transfer data between a computer and a mobile telephone, smartphone, BlackBerry, or other PDA (personal digital assistant).

Bluetooth is an FHSS system that splits the radio signal into tiny pieces. It moves among 79 different frequencies 1,600 times per second in the same unlicensed 2.4 GHz range as 802.11b and 802.11g Wi-Fi services. Bluetooth is not practical for connecting a computer to the Internet because it's slow (the maximum data transfer rate is only about 700Kbps), and it has a very limited signal range (most often about 33 feet, or 10 meters, or less).

In order to prevent interference between Bluetooth and Wi-Fi signals, many computers that use both technologies (including the widely used Intel Centrino chip set) coordinate the two services. When either module is active, it notifies the other, and the active service takes priority. This co-ordinated operation is slightly slower than either service operating alone, but the difference is insignificant.

Frequency Allocations

Each type of broadband wireless service uses a specific set of radio frequencies. Some of these frequencies are reserved for the exclusive use of a specific licensed service provider while others are free-for-all bands that are open for anybody to use.

Wi-Fi Services

The 802.11b, 802.11g, and 802.11n Wi-Fi services all operate in a frequency range at or slightly above 2.4 GHz. The 802.11a signal uses a band close to 5.3 GHz. The specific center frequencies of each Wi-Fi channel are listed in tables.

Unless you're a radio engineer, the important things to know about the different Wi-Fi services are the maximum data transmission rate and the signal range. Table shows the important characteristics of each Wi-Fi specification.

The differences between the maximum data speeds and the typical speeds are caused by the handshaking and other non-data information that must attach itself to each data packet. Obviously, there's a tremendous amount of overhead involved in moving information through any kind of Wi-Fi network.

Table: Wi-Fi Characteristics.

Type	Radio frequency	Signal range	Maximum data speed	Typical speed
802.11b	2.4Ghz	~ 30 meters (indoor) ~ 100 metres (outdoor)	11 Mbps	4Mbps
802.11a	5	~ 35 meters (indoor) ~ 110 metres (outdoor)	54 Mbps	23 Mbps
802.11g	2.4	~ 35 meters (indoor) ~ 110 metres (outdoor)	54 Mbps	20 Mbps
802.11n	2.4	~ 70 meters (indoor) ~ 160 metres (outdoor)	300 Mbps	120 Mbps

Other Broadband Services

National broadband wireless data service providers use a different variation on spread spectrum technology and a different range of radio frequencies. The broadband wireless services provided by Sprint, AT&T, and Verizon all share the frequencies around 800 MHz and 1,900 MHz used by those companies' digital mobile telephone networks. WiMAX services, such as Clearwire, use signals in the 2.3 to 2.5 GHz and 3.5 GHz bands.

Many new radio frequencies may open up for mobile telephone and data services in the United States after February 17, 2009, when all the existing analog television stations move to new digital channels, and the old VHF channels will close down. The newly vacant radio spectrum will become available for new services, including broadband wireless data.

The exact frequencies that the WiMAX and broadband wireless data services use are less important to you as a user than the frequencies of Wi-Fi signals because service providers control the base stations and access points. The network interface device in your computer automatically finds the right signal and sets up the connection without forcing you to choose a specific channel.

Choosing a Service

Each type of wireless access to the Internet offers a different combination of cost, coverage areas, reliability, ease of use, and security. Your choice will depend on your particular needs and the availability of signals in the locations where you need wireless Internet access.

For example, if you use your computer in just a few places and all of those places are within range of Wi-Fi hot spots, the built-in Wi-Fi adapter (or an inexpensive plug-in adapter) is probably your best choice. It's likely that Wi-Fi hot spots already exist at your workplace and in the libraries, coffee shops, schools, and conference centers where you regularly spend time, and it's relatively easy and inexpensive to install one or more access points at home. However, you will probably need a separate account to log in to each Wi-Fi network. Some of these Wi-Fi services are free, but others charge for access by the hour, by the day, or by the month; if you need paid accounts at several locations, the total cost can be more than a single account with a cellular service.

Wi-Fi also allows you to add portable computers to an existing LAN at home or in your school or workplace. And if cost is a primary concern, you will probably choose to use free public Wi-Fi hot spots instead of a cellular or WiMAX service that charges a monthly fee.

On the other hand, if you want constant Internet access wherever you go, the cellular data and WiMAX metropolitan area network services are better choices. Both systems provide coverage throughout large geographic regions, and both allow you to maintain a connection while you're moving from one place to another. You can use the same account and the same login and password every time you set up a connection. However, it's important to make sure that there's a usable cellular or WiMAX signal in all the places where you expect to use them before you commit to a long-term contract. Most wireless data service providers offer a free or low cost trial period that you can use to test the system.

As WiMAX and cellular data services become more common, many laptop computers and add-on network adapters will operate with both types of wireless services. When the computer detects a high-speed Wi-Fi signal, it will automatically try to establish a connection to that network. But when there is no local Wi-Fi signal, or if you haven't configured your computer to use any of the local signals, it will automatically shift over to your WiMAX or cellular data account and use that service to connect to the Internet.

All three types of wireless Internet services—Wi-Fi, cellular, and WiMAX—offer fast and reliable connections, but each has a different set of strengths and weaknesses. For short-range coverage and for access to local area networks, Wi-Fi is the obvious choice. If you are outside of the service areas of a DSL or cable Internet service, WiMAX is a huge improvement over a slow dial-up service. But when you carry your computer to many locations, a single account with a cellular or WiMAX service will allow you to connect to the Internet without the need to search for a new hot spot and set up a new account.

Wireless PAN

A Personal Area Network (PAN) refers to the connection of devices or equipment's in an individual user's environment typically within 10 meters or 33 feet. PAN is a network that enables people who sit at their comfortable premises to connect and interact with each other. It is the networking that plays a key role in a link and starts communications between hundreds and thousands.

PAN is an abbreviation for the Personal Area Network. PAN refers to a network that links devices within a range of 10 meters. The connection is possible with wired or wireless connections. The Wireless Personal Area Network (WPAN) is a small geographical network that responds to the needs of the individual user.

One or a few other computers, devices, telephones, video game consoles, or other electronic devices are included in a PAN network. However, if several users connect their devices in the same house, the Home Area Network or HAN network is more popular than PAN. The wired connection to a modem, which is further linked to multiple devices by either cable or cable, is given in this network.

HAN and PAN are both operated from one computer, but accessible across the network from all devices. Because of its versatility and capacity, PAN can be very helpful in offices and small network areas.

Wireless Personal Area Network

The Wireless Personal Area Network (WPAN), if interfaced with computers, could carry out extremely efficient activities. The Wireless Personal Area Network relies on the link between many devices. Once remote devices meet each other within a few meters, a WPAN offers a seamless and hassle-free connection.

The development and advancement of these innovations contribute to the effective running and functionality of consumer apps. A clear framework is established between all office and home devices by further development.

Wireless networks reach up to a height of 10 m (33 feet) with a few centimeters. These networks are a different category (or subset) of local area networks, which help an individual rather than a group. Secondary devices can attach to a PAN and use a primary computer to operate data. The system could be up to 100 meters wide with Bluetooth. Under some circumstances, PANs can still access the Internet. For example, a PAN system may connect to a LAN that has Internet connectivity and is itself a WAN. Important characteristics about PAN Network:

- Its range is generally short. The coverage gap is between 10 m and 100 m (Zigbee & Bluetooth) in a Wireless PAN network.
- Bluetooth is used for data transmission between devices, primarily used for low-data rate applications in home automation.
- PAN has the architecture of star and mesh.
- Zigbee, Bluetooth, Zwave, etc. PAN systems are used for low data rate and short-distance applications.
- In Zigbee PAN support 250kbps and Bluetooth support from Kbps to 24Mbps.
- In IoT (Internet of Things) networks, PAN wireless technologies have become quite popular.
- Devices in one PAN Network will connect to another PAN Network device in the range. PAN Network device in the range.

Technologies Used to Construct a PAN

Often USB (Universal Serial Bus) and fire-wire are connected to the PAN network together. This network can be wireless and can be connected by cables. A PAN can be small, wearable, and embedded devices, which communicate with other wireless devices during close contact. Bluetooth or infrared interfaces are used for WPANs.

Bluetooth WPAN

Bluetooth uses short-range waves for communication. Use within the WPAN includes tools for pointing, keyboards, and headsets of audio and video, printers, mobile phones, and computers. Also known as the piconet is a WPAN Bluetooth. It consists of a total of 8 active devices that are connected with the master-slave system.

The first piconet is the master-slave relationship in the network. The Bluetooth long-range routers have a reach of up to 1000 ft. In a mesh network, we can also use this network. In this way, you can move the data from one component to another.

Infrared Data Association (IrDA)

Infrared technology is used in the Infrared Data Association. Infrared is used for remote controls in general. Many other devices use IrDA like printers, keyboards, etc. The limits of scope for WPANs vary in size from centimeters to a few meters.

Examples of Wireless Personal Area Network (WPAN)

Some of the examples of Wireless Personal area network are as follows:

- Wireless keyboards,
- Wireless printers,
- Gaming consoles,
- TV remotes,
- Bluetooth,
- Smartphones,
- Wireless USB,
- Wireless Mouse.

Advantages and Disadvantages of PAN

PANs are for individual use so the benefits are more easily understood than for large area structures such as WLAN and MAN. Below are some of the advantages of PAN:

- Wireless networks save customers from the burden of cable management, often complicated and unwanted floor space. So no extra space is required.
- Bluetooth PAN allows users to conveniently control their computers and systems so that errors can be repaired quickly.
- Fast wireless system access and control to boost performance. For instance, a business can send and collect information to its staff using intelligent devices, all with a single PAN.
- Easy and efficient Bluetooth-enabled computer data synchronization and communication.
- With a wireless PAN link, the use of innovative new technology is easily made to help the client.
- PAN gives its users cost savings and performance.
- It is easier to repair bugs and faults in a PAN connection due to its easy structure.
- Bluetooth PANs provide its users with far more security.
- It is portable.

Disadvantages of PAN are as follows:

- PAN uses a short-range Bluetooth communication that is capable of supporting only a few devices nearby.

- Often PANs can communicate on the same bandwidth with other wireless devices and interrupt the communication.
- Due to Bluetooth connections, PANs operate slower than other networks.

Wireless LAN

Wireless LAN stands for Wireless Local Area Network. It is also called LAWN (Local Area Wireless Network). WLAN is one in which a mobile user can connect to a Local Area Network (LAN) through a wireless connection.

The IEEE 802.11 group of standards defines the technologies for wireless LANs. For path sharing, 802.11 standard uses the Ethernet protocol and CSMA/CA (carrier sense multiple access with collision avoidance). It also uses an encryption method i.e. wired equivalent privacy algorithm.

Wireless LANs provide high speed data communication in small areas such as building or an office. WLANs allow users to move around in a confined area while they are still connected to the network.

In some instance wireless LAN technology is used to save costs and avoid laying cable, while in other cases, it is the only option for providing high-speed internet access to the public. Whatever the reason, wireless solutions are popping up everywhere.

Advantages of WLANs

- Flexibility: Within radio coverage, nodes can communicate without further restriction. Radio waves can penetrate walls, senders and receivers can be placed anywhere (also non-visible, e.g., within devices, in walls etc).
- Planning: Only wireless ad-hoc networks allow for communication without previous planning, any wired network needs wiring plans.
- Design: Wireless networks allow for the design of independent, small devices which can for example be put into a pocket. Cables not only restrict users but also designers of small notepads, PDAs, etc.
- Robustness: Wireless networks can handle disasters, e.g., earthquakes, flood etc. whereas, networks requiring a wired infrastructure will usually break down completely in disasters.
- Cost: The cost of installing and maintaining a wireless LAN is on average lower than the cost of installing and maintaining a traditional wired LAN, for two reasons. First, after providing wireless access to the wireless network via an access point for the first user, adding additional users to a network will not increase the cost. And second, wireless LAN eliminates the direct costs of cabling and the labor associated with installing and repairing it.
- Ease of Use: Wireless LAN is easy to use and the users need very little new information to take advantage of WLANs.

Disadvantages of WLANs

- Quality of Services: Quality of wireless LAN is typically lower than wired networks. The main reason for this is the lower bandwidth due to limitations is radio transmission, higher error rates due to interference and higher delay/delay variation due to extensive error correction and detection mechanisms.
- Proprietary Solutions: Due to slow standardization procedures, many companies have come up with proprietary solutions offering standardization functionality plus many enhanced features. Most components today adhere to the basic standards IEEE 802.11a or 802.11b.
- Restrictions: Several govt. and non-govt. institutions world-wide regulate the operation and restrict frequencies to minimize interference.
- Global Operation: Wireless LAN products are sold in all countries so, national and international frequency regulations have to be considered.
- Low Power: Devices communicating via a wireless LAN are typically power consuming, also wireless devices running on battery power. Whereas the LAN design should take this into account and implement special power saving modes and power management functions.
- License Free Operation: LAN operators don't want to apply for a special license to be able to use the product. The equipment must operate in a license free band, such as the 2.4 GHz ISM band.
- Robust Transmission Technology**:** If wireless LAN uses radio transmission, many other electrical devices can interfere with them (such as vacuum cleaner, train engines, hair dryers, etc.).Wireless LAN transceivers cannot be adjusted for perfect transmission is a standard office or production environment.

Types of Wireless LAN

Peer-to-Peer

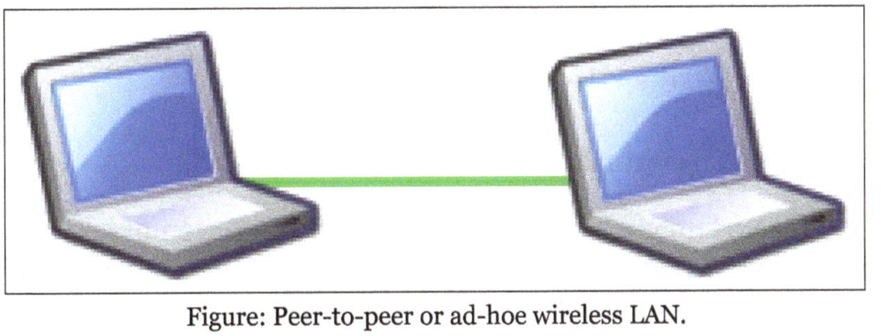

Figure: Peer-to-peer or ad-hoe wireless LAN.

An ad-hoc network is a network where stations communicate only peer-to-peer (P2P). There is no base and no one gives permission to talk. This is accomplished using the Independent Basic Service Set (IBSS). A peer-to-peer (P2P) network allows wireless devices to directly communicate with each other. Wireless devices within range of each other can discover and communicate directly without involving central access points. This method is typically used by two computers so that they can connect to each other to form a network. If a signal strength meter is used in this situation, it may not read the strength accurately and can be misleading, because it registers the strength of the strongest signal, which may be the closest computer.

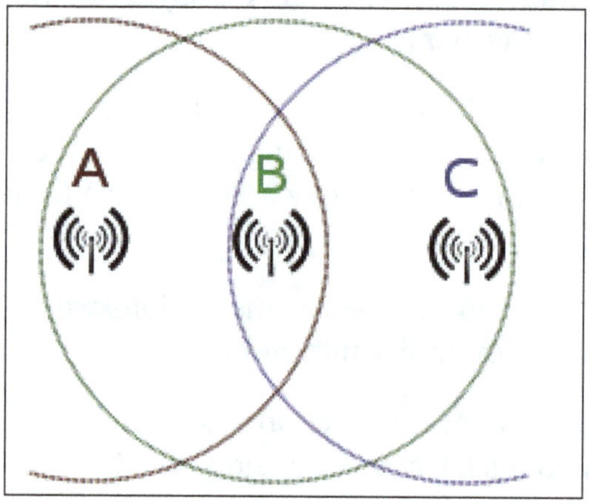

Hidden node problem: Devices A and C are both communicating with B, but are unaware of each other IEEE 802.11 defines the physical layer (PHY) and MAC (Media Access Control) layers based on CSMA/CA (Carrier Sense Multiple Access with Collision Avoidance). The 802.11 specification includes provisions designed to minimize collisions, because two mobile units may both be in range of a common access point, but out of range of each other.

The 802.11 has two basic modes of operation: Ad hoc mode enables peer-to-peer transmission between mobile units. Infrastructure mode in which mobile units communicate through an access point that serves as a bridge to a wired network infrastructure is the more common wireless LAN application the one being covered. Since wireless communication uses a more open medium for communication in comparison to wired LANs, the 802.11 designers also included shared-key encryption mechanisms: Wired Equivalent Privacy (WEP), Wi-Fi Protected Access (WPA, WPA2), to secure wireless computer networks.

Bridge

A bridge can be used to connect networks, typically of different types. A wireless Ethernet bridge allows the connection of devices on a wired Ethernet network to a wireless network. The bridge acts as the connection point to the Wireless LAN.

Wireless Distribution System

A Wireless Distribution System enables the wireless interconnection of access points in an IEEE 802.11 network. It allows a wireless network to be expanded using multiple access points without the need for a wired backbone to link them, as is traditionally required. The notable advantage of WDS over other solutions is that it preserves the MAC addresses of client packets across links between access points.

An access point can be either a main, relay or remote base station. A main base station is typically connected to the wired Ethernet. A relay base station relays data between remote base stations, wireless clients or other relay stations to either a main or another relay base station. A remote base station accepts connections from wireless clients and passes them to relay or main stations. Connections between "clients" are made using MAC addresses rather than by specifying IP assignments.

All base stations in a Wireless Distribution System must be configured to use the same radio channel, and share WEP keys or WPA keys if they are used. They can be configured to different service set identifiers. WDS also requires that every base station be configured to forward to others in the system. WDS may also be referred to as repeater mode because it appears to bridge and accept wireless clients at the same time (unlike traditional bridging). It should be noted; however, that throughput in this method is halved for all clients connected wirelessly. When it is difficult to connect all of the access points in a network by wires, it is also possible to put up access points as repeaters.

Roaming

Roaming among Wireless Local Area Networks

There are two definitions for wireless LAN roaming:

- Internal Roaming: The Mobile Station (MS) moves from one access point (AP) to another AP within a home network because the signal strength is too weak. An authentication server (RADIUS) presumes the re-authentication of MS via 802.1x (e.g. with PEAP). The billing of QoS is in the home network. A Mobile Station roaming from one access point to another often interrupts the flow of data among the Mobile Station and an application connected to the network. The Mobile Station, for instance, periodically monitors the presence of alternative access points (ones that will provide a better connection). At some point, based on proprietary mechanisms, the Mobile Station decides to re-associate with an access point having a stronger wireless signal. The Mobile Station, however, may lose a connection with an access point before associating with another access point. In order to provide reliable connections with applications, the Mobile Station must generally include software that provides session persistence.
- External Roaming: The MS (client) moves into a WLAN of another Wireless Internet Service Provider (WISP) and takes their services (Hotspot). The user can independently of his home network use another foreign network, if this is open for visitors. There must be special authentication and billing systems for mobile services in a foreign network.

IEEE 802.11 is a set of standards for implementing wireless local area network (WLAN) computer communication in the 2.4, 3.6 and 5 GHz frequency bands. They are created and maintained by the IEEE LAN/MAN Standards Committee (IEEE 802). The base version of the standard IEEE 802.11-2007 has had subsequent amendments. These standards provide the basis for wireless network products using the Wi-Fi brand name.

The 802.11 family consists of a series of over-the-air modulation techniques that use the same basic protocol. The most popular are those defined by the 802.11b and 802.11g protocols, which are amendments to the original standard. 802.11-1997 was the first wireless networking standard, but 802.11b was the first widely accepted one, followed by 802.11g and 802.11n. 802.11n is a new multi-streaming modulation technique.

802.11b and 802.11g use the 2.4 GHz ISM band, operating in the United States under Part 15 of the US Federal Communications Commission Rules and Regulations. Because of this choice of frequency band, 802.11b and g equipment may occasionally suffer interference from microwave ovens, cordless telephones and Bluetooth devices. 802.11b and 802.11g control their interference and susceptibility to interference by using direct-sequence spread spectrum (DSSS) and orthogonal frequency-division multiplexing (OFDM) signaling methods, respectively. 802.11a uses the 5 GHz U-NII band, which, for much of the world, offers at least 23 non-overlapping channels rather than the 2.4 GHz ISM frequency band, where adjacent channels overlap. Better or worse performance with higher or lower frequencies (channels) may be realized, depending on the environment.

The segment of the radio frequency spectrum used by 802.11 varies between countries. In the US, 802.11a and 802.11g devices may be operated without a license, as allowed in Part 15 of the FCC Rules and Regulations. Frequencies used by channels one through six of 802.11b and 802.11g fall within the 2.4 GHz amateur radio band. Licensed amateur radio operators may operate 802.11b/g devices under Part 97 of the FCC Rules and Regulations, allowing increased power output but not commercial content or encryption.

Protocols

- A1 A2 IEEE 802.11y-2008 extended operation of 802.11a to the licensed 3.7 GHz band. Increased power limits allow a range up to 5,000 m. As of 2009, it is only being licensed in the United States by the FCC.
- B1 B2 Assumes short guard interval (SGI) enabled otherwise reduces each data rate by 10%.

Table: 802.11 network standards.

802.11 Protocol	Release	Freq. (Ghz)	Bandwidth (Mhz)	Data Rate Per Stream (Mbits/S)	Allowable Mimo Streams	Modulation	Approx. Indoor Range	Approx. Outdoor Range
	Jun 1997	2.4 5	20	1,2 6,9,12,18	1	Dsss, Fhss	20	100

A	Sep 1999	3.7	20	24,36,48, 54	1	Odfm, Qafm	35	120
B	Sep 1999	2.4	20	5.5,11	1	Dsss	-	5,000
G	Jun 2003	2.4	20 20	6,9,12,18, 24, 36, 48, 54 7.2,14.4 21.7,28.9 43.3, 57.8, 65, 72.2	1 1	Odfm Qam, Dsss	38 38	140 140
N	Oct 2009	2.41/5	40 80	15,30,45 60,90,120 135,150 433,150	4	Odfm Qam	70 70	250 250
Ac	Nov 11	5	160	867,1.73 Gbit/S,3.47 Gbit/S,6.93 Gbit/S	8	Odfm Qam		

802.11-1997 (802.11 Legacy)

The original version of the standard IEEE 802.11 was released in 1997 and clarified in 1999, but is today obsolete. It specified two net bit rates of 1 or 2 megabits per second (Mbit/s), plus forward error correction code. It specified three alternative physical layer technologies: diffuse infrared operating at 1 Mbit/s; frequency-hopping spread spectrum operating at 1 Mbit/s or 2 Mbit/s; and direct-sequence spread spectrum operating at 1 Mbit/s or 2 Mbit/s. The latter two radio technologies used microwave transmission over the Industrial Scientific Medical frequency band at 2.4 GHz. Some earlier WLAN technologies used lower frequencies, such as the U.S. 900 MHz ISM band. Legacy 802.11 with direct-sequence spread spectrum was rapidly supplanted and popularized by 802.11b.

802.11a

The 802.11a standard uses the same data link layer protocol and frame format as the original standard, but an OFDM based air interface (physical layer). It operates in the 5 GHz band with a maximum net data rate of 54 Mbit/s, plus error correction code, which yields realistic net achievable throughput in the mid-20 Mbit/s.

Since the 2.4 GHz band is heavily used to the point of being crowded, using the relatively unused 5 GHz band gives 802.11a a significant advantage. However, this high carrier frequency also brings a disadvantage: the effective overall range of 802.11a is less than that of 802.11b/g. In theory,

802.11a signals are absorbed more readily by walls and other solid objects in their path due to their smaller wavelength and, as a result, cannot penetrate as far as those of 802.11b. In practice, 802.11b typically has a higher range at low speeds (802.11b will reduce speed to 5 Mbit/s or even 1 Mbit/s at low signal strengths). 802.11a too suffers from interference, but locally there may be fewer signals to interfere with, resulting in less interference and better throughput.

802.11b

802.11b has a maximum raw data rate of 11 Mbit/s and uses the same media access method defined in the original standard. 802.11b products appeared on the market in early 2000, since 802.11b is a direct extension of the modulation technique defined in the original standard. The dramatic increase in throughput of 802.11b (compared to the original standard) along with simultaneous substantial price reductions led to the rapid acceptance of 802.11b as the definitive wireless LAN technology. 802.11b devices suffer interference from other products operating in the 2.4 GHz band. Devices operating in the 2.4 GHz range include: microwave ovens, Bluetooth devices, baby monitors, and cordless telephones.

802.11g

In June 2003, a third modulation standard was ratified: 802.11g. This works in the 2.4 GHz band (like 802.11b), but uses the same OFDM based transmission scheme as 802.11a. It operates at a maximum physical layer bit rate of 54 Mbit/s exclusive of forward error correction codes, or about 22 Mbit/s average throughputs. 802.11g hardware is fully backwards compatible with 802.11b hardware and therefore is encumbered with legacy issues that reduce throughput when compared to 802.11a by ~21%.

The then-proposed 802.11g standard was rapidly adopted by consumers starting in January 2003, well before ratification, due to the desire for higher data rates as well as to reductions in manufacturing costs. By summer 2003, most dual-band 802.11a/b products became dual band/tri-mode, supporting a and b/g in a single mobile adapter card or access point. Details of making b and g work well together occupied much of the lingering technical process; in an 802.11g network, however, activity of an 802.11b participant will reduce the data rate of the overall 802.11g network. Like 802.11b, 802.11g devices suffer interference from other products operating in the 2.4 GHz band, for example wireless keyboards.

802.11-2007

In 2003, task group TGma was authorized to “roll up” many of the amendments to the 1999 version of the 802.11 standard. REVma or 802.11ma, as it was called, created a single document that merged 8 amendments (802.11a, b, d, e, g, h, i, j) with the base standard. Upon approval on March 8, 2007, 802.11REVma was renamed to the then-current base standard IEEE 802.11- 2007.

802.11n

802.11n is an amendment which improves upon the previous 802.11 standards by adding multiple-input multiple-output antennas (MIMO). 802.11n operates on both the 2.4 GHz and the lesser used 5 GHz bands. Prior to the final ratification, enterprises was already migrating to 802.11n networks based on the Wi-Fi Alliance's certification of products conforming to a 2007 draft of the 802.11n proposal.

802.11ac

IEEE 802.11ac is a standard under development which will provide high throughput in the 5 GHz band. This specification will enable multi-station WLAN throughput of at least 1 Gigabit per second and a maximum single link throughput of at least 500 megabit per second, by using wider RF bandwidth, more l streams (up to 8), and high-density modulation (up to 256 QAM).

Difference between PAN and LAN

Feature	PAN	LAN
Definition	It's a network, normally, covering a very small area. This helps computing devices to connect and share information and data with other nearby computers.	The network connects computers within a limited distance, such as a house or a single lab made up of several computers.
Data Transfer	Bluetooth is used for the transfer of data between devices, especially for low-data-rate applications in home automation.	It is designed to transfer a high data rate from Mbps to Gbps.
Application	Designed for devices with low data rates and short distances.	It is mainly used in LAN and LAN wireless applications, where high-speed data transfer is required.
Range	The coverage distance in a wireless PAN (Zigbee& Bluetooth) network is between 10 m and 100 m.	The scope is usually between buildings or within a single building. The coverage limit is 10 m to 100 m or higher in case of wireless LAN.
Speed	PAN supports 250kbps in Zigbee, Bluetooth support from Kbps to 24Mbps.	10, 100, and 1000 Mbps are supported by LAN.

Wireless Ad-hoc Network

A mobile Ad-hoc network is a self-configuring infrastructure less network of mobile devices connected by wireless links. Where no centralize authority is available and all nodes are independently follow the routine of mobility. Due to absence of the centralize authority the attacker easily join the network and perform the malicious activity. Beyond this gap to overcome this fault or problem various trust based security architecture are proposed for MANET. In general two types of communications can be considered in classical MANETs, broadcast communications and multi-hop communications via routing protocols.

The mobile ad hoc network has typical features namely Unreliability of wireless links between nodes and constantly changing topology. Each device in a MANET is free to move independently in any direction, and will therefore change its links to other devices frequently. Each must forward traffic unrelated to its own use, and therefore be a router. The primary challenge in building a MANET is equipping each device to continuously maintain the information required to properly route traffic. Such networks may operate by themselves or may be connected to the larger Internet. The achievement of MANET is hug growth of laptops and wireless or Wi-Fi networking.

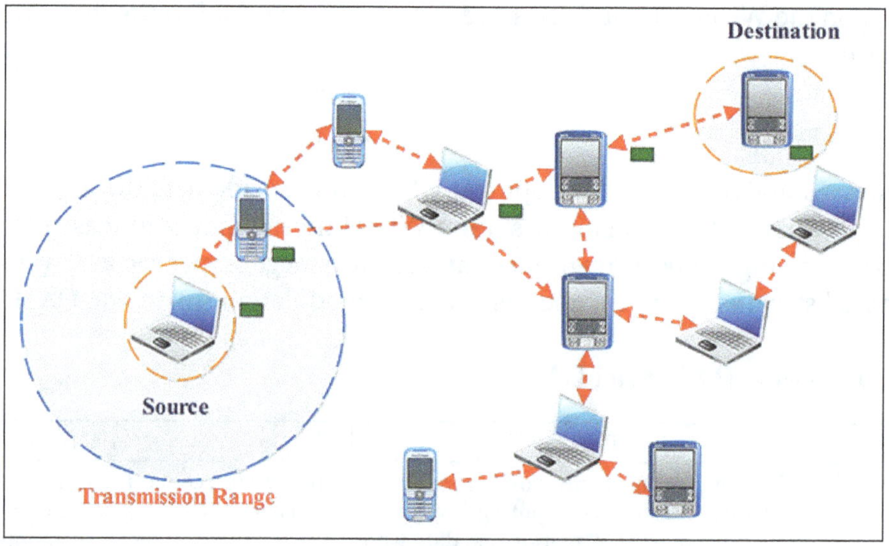

Figure: Mobile Ad Hoc Network.

Types of MANET

Mobile Ad Hoc Networks are further classified into following three types:

Vehicular Ad-hoc Network (VANET)

Vehicular Ad-hoc Network (VANET), a subclass of mobile Ad Hoc networks (MANETs). Vehicular Ad Hoc Networks is a class of special wireless ad hoc network with the characteristics of high node mobility and fast topology changes. A Vehicular Ad-Hoc Network is a technology that uses moving cars as nodes in a network to create a mobile network. VANET turns every participating car into a wireless router or node, allowing cars approximately 100 to 300 metres of each other to connect and, in turn, create a network with a wide range. As cars fall out of the signal range and drop out of the network, other cars can join in, connecting vehicles to one another so that a mobile Internet is created. It is a promising approach for future intelligent transportation system (ITS). These networks have no fixed infrastructure and instead rely on the vehicles themselves to provide network functionality. However, due to mobility constraints, driver behavior, and high mobility, VANETs exhibit characteristics that are dramatically different from many generic MANETs.

Internet based Mobile Ad-hoc Network (IMANET)

Internet Based Mobile Ad-hoc Networks are ad hoc networks that link mobile nodes and fixed Internet gateway nodes. In such type of networks normal ad hoc routing algorithms don't apply directly. Internet Based Mobile Ad-hoc Networks are ad-hoc networks that link mobile nodes and fixed Internet-gateway nodes. In such type of networks normal ad hoc routing algorithms don't apply directly. Wireless networks can generally be classified as wireless fixed networks, and wireless, or mobile ad-hoc networks. MANETs (mobile ad-hoc networks) are based on the idea of establishing a network without taking any support from a centralized structure. By nature these types of networks are suitable for situations where either no fixed infrastructure exists, or to deploy one is not possible.

Intelligent Vehicular Ad-hoc Network (InVANET)

Intelligent Vehicular Ad-hoc Network is a kind of artificial intelligence that helps vehicles to behave in intelligent manners during vehicle-to-vehicle collisions, accidents, drunken driving etc. Intelligent Vehicular Ad-hoc Networking defines an intelligent way of using Vehicular Networking. INVANET integrates on multiple ad-hoc networking technologies such as Wi-Fi IEEE 802.11, WAVE IEEE 1609, WiMAX, IEEE 802.16, Bluetooth, IRA, and ZigBee for easy, accurate, effective and simple communication between vehicles on dynamic mobility. Effective measures such as media communication between vehicles can be enabled as well methods to track the automotive vehicles are also preferred. InVANET helps in defining safety measures in vehicles, streaming communication between vehicles, infotainment and telematics.

Applications of Mobile Ad-hoc Network

Some of the typical applications include:

- Collaborative work: For some business environments, the need for collaborative computing might be more important outside office environments than inside and where people do need to have outside meetings to cooperate and exchange information on a given project.
- Military battlefield: Ad-hoc networking would allow the military to take advantage of commonplace network technology to maintain an information network between the soldiers, vehicles, and military information head quarter.
- Local level: Ad-hoc networks can autonomously link an instant and temporary multimedia network using notebook computers to spread and share information among participants at e.g. conference or classroom. Another appropriate local level application might be in home networks where devices can communicate directly to exchange information.
- Personal area network and Bluetooth: A personal area network is a short range, localized network where nodes are usually associated with a given person. Short-range MANET such as Bluetooth can simplify the inter communication between various mobile devices such as a laptop, and a mobile phone.
- Commercial sector: Ad-hoc can be used in emergency/rescue operations for disaster relief efforts, e.g. in fire, flood, or earthquake.

Wireless MAN

Wireless MANs offer connections between buildings and users within a city or campus area through several system configurations. In most cases, the wireless MAN beams RF or infrared light from one point to other using directive antennae.

Point-to-point Systems

A point-to-point solution uses RF or infrared signals that utilize either semi directional or highly directional antennae to extend range across metropolitan areas, such as college campuses and cities. Range can be as high as 30 miles for RF systems using highly directional antennae. Figure illustrates a point-to-point wireless MAN system.

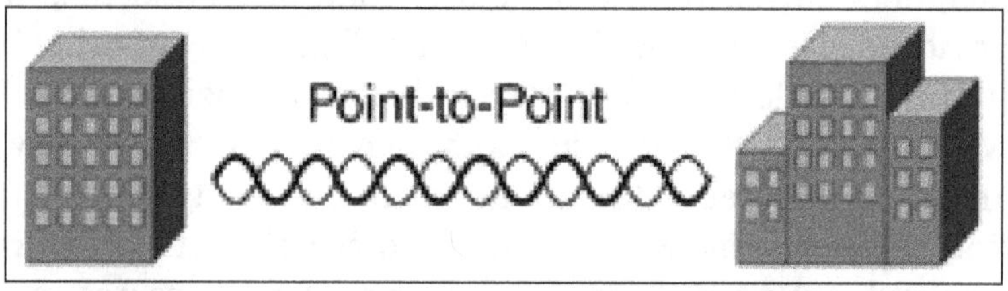

Figure: Point-to-Point Wireless MAN Directly Connects Two Points in the Network.

A medical center, for example, can use a point-to-point wireless MAN to provide a communications link between the main hospital and a remote clinic within the same city. This resulting system, however, does not provide as much flexibility as point-to-multipoint solutions. However, if there is a need to connect only a couple sites, the cost of implementing a point-to-point system is less compared to a point-to-multipoint system.

Point-to-Multipoint System

A typical point-to-multipoint link utilizes a centralized omnidirectional antenna that provides a single transceiver point for tying together multiple remote stations. For example, a building within the center of a city can host the omnidirectional antenna, and other nearby metropolitan-area buildings can point directional antennae at the centralized location. The central transceiver receives and retransmits the signals.

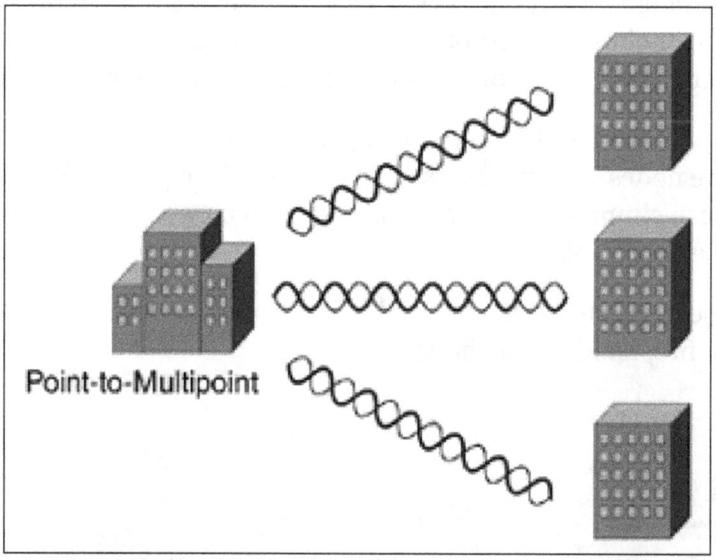

Figure: Point-to-Multipoint Wireless MAN Interconnects Users Through a Common, Centralized Transceiver.

A strong advantage of the point-to-multipoint wireless MAN is that it makes the addition of new connections easy. In fact, this approach can be less expensive compared to point-to-point systems when there are multiple sites to interconnect or connect to a central location. For example, a company headquarters having many remote warehouses and manufacturing plants within the same city or rural area would benefit from a point-to-multipoint system.

Packet Radio Systems

A packet radio system utilizes special wireless routers that forward data contained within packets to the destination. Each user has a packet radio NIC that transmits data to the nearest wireless router. This router then retransmits the data to the next router. This hopping from router to router occurs until the packet reaches the destination. This mesh type networking is not new. Amateur Ham radio operators have used it for decades, and companies such as Metricom have been deploying these types of systems in cities for nearly 10 years.

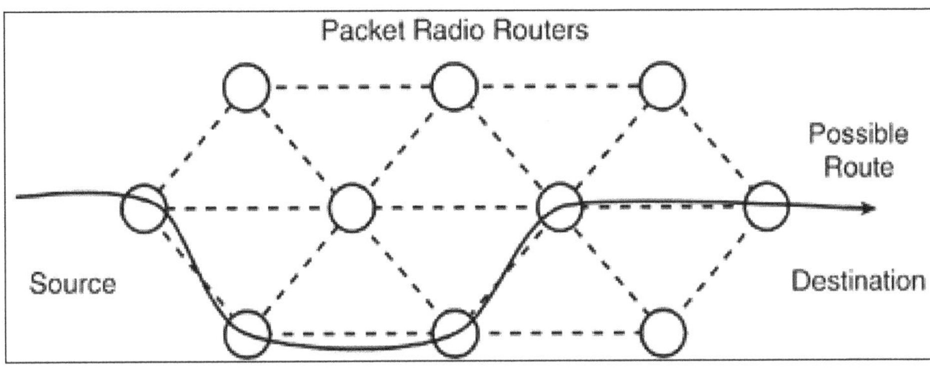

Figure: Packet Radio System Hops Data Packets from the Source to Destination.

A city government might want to deploy a packet radio system to offer wireless connectivity for supporting applications through the entire city area. The installation of routers in strategic places through the city provides the necessary infrastructure. There's no need for wires for interconnecting the routers.

This form of networking is also survivable. If one router becomes inoperative, perhaps because of a lightning strike or sabotage, adaptive routing protocols automatically update routing tables in each router so that data packets will avoid traversing the inoperative router.

Wireless WAN

A wireless Wide Area Network (WWAN) is a computer data network that may extend over a large geographical area. Characteristically, a WWAN receives and transmits data using radio signals over the initial interconnect with the mobile computer system. At the switching center, WWAN segments then connect to either a specialized public or private network via telephone or other high-speed communication links. These back-hauls interconnects then link to an organization's existing LAN/ WAN infrastructure. Recent developments now allow direct connections to generic public networks, like the Internet, further reducing the cost of deployment.

Unlike wireless local area networks (WLANs), the coverage area for WWANs is normally measured in miles rather than feet. Data throughput rates for WWANs are a great deal slower than they are for local area network connections. They are typically measured in tens of thousands of bytes per second or slower. The average access time may be slower than a 28.8 modem connection. Wireless networks in general are also more susceptible to environmental factors, like weather and terrain, than wired networks.

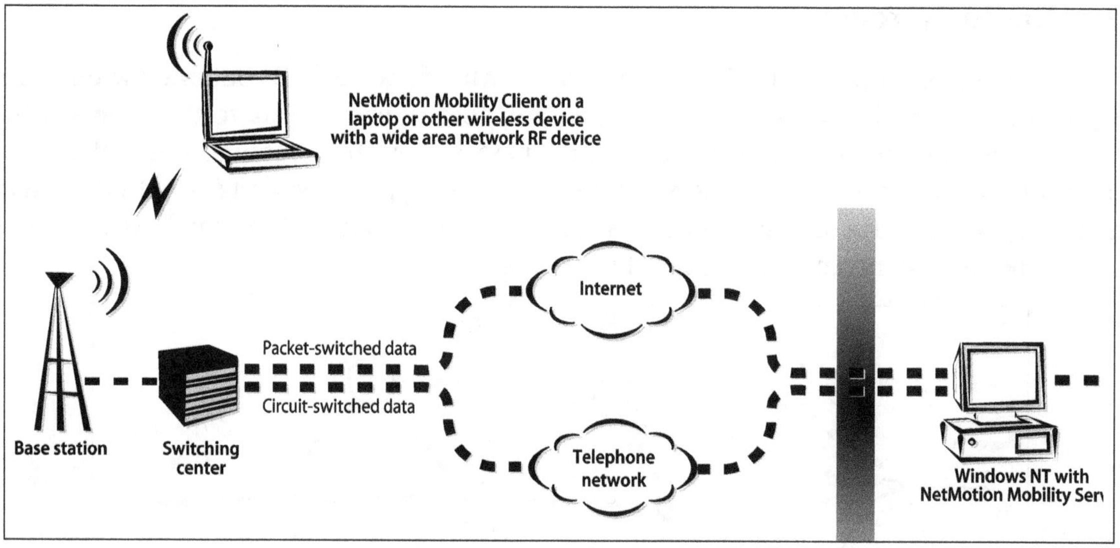

Components of a wireless WAN.

Wireless WAN Benefits

Whether they are in the office or traveling, today's professionals need access to the corporate data and applications that keep them connected to their organization's processes. Wireless WAN connectivity means hospitals can track patients and their charts remotely. Sales staff has access to the latest product information for customer presentations. Managers can get the latest data to make decisions when they are away from their offices. Or public safety officials can access data that they need from the field.

It is access to data communications connections wherever and whenever they are needed that is driving the market. Cahners In-Stat's numbers, while lower (2 million users in 1999 growing to 25 million in 2003), echo similar growth. Both agree that in the near term, it is corporate users who will drive this growth. This business demand will eventually lower future wireless access costs and open the doors to the consumer market.

Wireless WAN Limitations

Mobile computing devices are getting smaller, cheaper and more powerful. At the same time the amount of information available today is growing astronomically. The demand for connecting mobile devices to content-rich networks is also rising quickly, and it would seem that WWAN technology is the perfect answer. But today's wireless WANs have limitations in several areas, including:

- Security,
- Performance,
- Application persistence,
- Roaming,
- Off-the-shelf solutions,
- Central management.

To overcome these limitations, organizations usually have to increase their already steep IT workload by developing point solutions.

Security

In modern network topologies, physical boundaries between public and private networks are blurring. Unless care is taken, when a company decides to add wireless connectivity to this mix, their information security can be compromised. Wireless systems in general broadcast data over radio waves into the ether. The security implications are obvious—what companies broadcast goes everywhere. One of the first concerns that the IT department has is whether the information crossing the wireless network boundary is secure, because it is susceptible to eavesdropping. Secondly, they must feel confident that a user has the necessary network permissions to access information, since they cannot rely on the mobile device's physical location to tell them. Wirelessly connected devices can easily breech two of the most fundamental network security policies: authentication and confidentiality.

Performance

Things never encountered in wired infrastructures affect wireless connections: environmental factors, such as user movement, harmonics, terrain, and weather, can easily affect throughput and performance. Today's standard network topologies and protocol implementations were designed for communications between stationary systems. In general, when a mobile device communicates with a stationary device, the stationary system uses the same algorithms to communicate with its mobile peer. It is not aware that the device is mobile, and it is this lack of awareness that is at the heart of most performance problems. Application performance can further degrade when generic transport implementations intended for stationary devices are deployed on mobile ones.

Another concern is how these generic solutions deal with user movement across network segments. Since mobility was never considered when most networks were originally designed, additional information is required at the protocol layer. This additional overhead can further reduce overall performance and provide less than optimal throughput over wireless connections.

Application Persistence

Because users with wireless devices are not tethered to a fixed interconnect, they can unwittingly move out of a coverage area. They may also choose to suspend operations to extend battery life. Such normal communication interruptions can wreak havoc with existing transport protocol implementations and cause user applications to either lock up or terminate.

To minimize such interruptions, most wireless solutions require customizing both the applications and the network. As a result, most current solutions are limited in scope and restricted to a specific set of applications. The mission-critical applications are still relegated to stationary or desktop systems. When a company tries to deploy these applications to their mobile devices, mobile users spend a lot of time getting back on the network after losing their connections. They must re-login, re-authenticate, possibly navigate through multiple screens to find the point where they left off, and then re-enter any lost data. In the end, users become frustrated and feel unproductive; this costs the company precious time and money.

Roaming

Over the past several years, network installations have been segmented for management and security reasons. To aid network segmentation deployment, the networking industry has worked hard to allow dynamic configuration of systems when they attach to a particular segment. However, once application sessions are established over existing protocols, the network assumes that a device remains attached to that segment for the duration of the session. But mobile devices can wander to different areas and the communications pipe can be easily broken. To avoid this, companies have had to either re-architect their network interconnections or customize their applications, both of which cost time and money.

Standard off-the-shelf Applications

Most companies do not own the source code for applications that they use to access their mission-critical information, especially the networking components. They either purchase canned applications or libraries that enable access to the stored information. If the latter is the case, these libraries (in binary form) are then linked in to customized front-end applications. These "off-the-shelf" applications may suffer from the mobility problems mentioned above. To re-customize each important application for mobile deployment strains, the IT department's personnel resources and increases the time and cost of deployment. Once completed, the newly customized application may require retraining and extend the deployment timeline because of a complex installation.

Centralized Management and Efficient use of IT Resources

Troubleshooting a network problem is already a difficult task for IT professionals. Most of the devices are stationary and most of the tools help pinpoint a problem to a specific physical location. Mobile devices only increase the troubleshooting complexity for IT professionals since there is no physical connection and difficult network problems can be increasingly intermittent. Troubleshooting errant mobile devices can make even the most experienced network troubleshooter armed with the most sophisticated tools still feel like a novice.

As IT resources are spread thinner and thinner, companies need to consider how this additional load can be added onto an already overworked staff. No wonder this is one of the major barriers that organizations must overcome when deploying wireless communications technologies.

Cellular Network

Mobile networks are also known as *cellular networks*. They're made up of "cells," which are areas of land that are typically hexagonal, have at least one transceiver cell tower within their area, and use various radio frequencies. These cells connect to one another and to telephone switches or exchanges. Cell towers connect to each other to hand off packets of signals—data, voice, and text messaging—ultimately bringing these signals to mobile devices such as phones and tablets that act as receivers. Providers use each other's' towers in many areas, creating a complex web that offers the widest possible network coverage to subscribers.

Frequencies

Many network subscribers use mobile networks' frequencies at the same time. Cell tower sites and mobile devices manipulate the frequencies so that they can use low-power transmitters to supply their services with the least possible interference.

3G, 4G and 5G Networks

Mobile networks have evolved through a series of generations, each representing significant technological improvements over the previous generations. The first two generations of mobile networks first introduced analog voice (1G) and then digital voice (2G). Subsequent generations supported the proliferation of smartphones by introducing data connections (3G) and allowing access to the internet. 4G service networks improved data connections, making them faster and better able to provide greater bandwidth for uses such as streaming.

The latest technology is the 5G network, which promises even faster speeds and greater bandwidth compared with 4G while reducing interference with other nearby wireless devices. Where 4G uses frequencies below 6 GHz, newer 5G networks use shorter wavelength signals with much higher frequencies, in the range of 30 GHz to 300 GHz. These frequencies provide higher bandwidth and allow signals to be more directional, thus reducing interference.

Leading Mobile Network Providers

Cellular service providers in the U.S. range in size from small, regional companies to large, well-known corporations in the telecommunications field, such as Verizon Wireless, AT&T, T-Mobile, US Cellular, and Sprint.

Types of Mobile Networks

The mobile technologies that large mobile service providers' use varies and mobile devices are built to use the technology of the intended carrier and region. The two main mobile technologies in use are Global System for Mobile communications, which is an international standard, and Code Division Multiple Access, owned by Qualcomm. GSM phones don't work on CDMA networks, and vice versa. Long-Term Evolution is based on GSM and offers greater network capacity and speed.

GSM Vs. CDMA Mobile Networks

Signal reception, call quality, and speed all depend on many factors. The user's location, service provider, and equipment all play a role. GSM and CDMA don't differ much on quality, but the way they work does.

From a consumer standpoint, GSM is more convenient because a GSM phone carries the entire customer's data on a removable SIM card; to change phones, the customer simply swaps the SIM card into the new GSM phone, and it connects to the provider's GSM network. A GSM network must accept any GSM-compliant phone, leaving consumers quite a bit of freedom over their choices in equipment.

CDMA phones, on the other hand, aren't as easily transferred between carriers. CDMA carriers identify subscribers based on safe lists, not SIM cards, and only approved phones are allowed on their networks. Some CDMA phones use SIM cards, but these are for the purpose of connecting to LTE networks or for flexibility when the phone is used outside of the U.S.

Features of Cellular Systems

Wireless Cellular Systems solves the problem of spectral congestion and increases user capacity. The features of cellular systems are as follows:

- Offer very high capacity in a limited spectrum.
- Reuse of radio channel in different cells.
- Enable a fixed number of channels to serve an arbitrarily large number of users by reusing the channel throughout the coverage region.
- Communication is always between mobile and base station (not directly between mobiles).
- Each cellular base station is allocated a group of radio channels within a small geographic area called a cell.
- Neighboring cells are assigned different channel groups.
- By limiting the coverage area to within the boundary of the cell, the channel groups may be reused to cover different cells.
- Keep interference levels within tolerable limits.
- Frequency reuse or frequency planning.
- Organization of Wireless Cellular Network.

Cellular network is organized into multiple low power transmitters each 100w or less.

Shape of Cells

The coverage area of cellular networks is divided into cells, each cell having its own antenna for transmitting the signals. Each cell has its own frequencies. Data communication in cellular networks is served by its base station transmitter, receiver and its control unit.

The shape of cells can be either square or hexagon:

Square

A square cell has four neighbors at distance d and four at distance Root 2 d:

- Better if all adjacent antennas equidistant.
- Simplifies choosing and switching to new antenna.

Hexagon

A hexagon cell shape is highly recommended for its easy coverage and calculations. It offers the following advantages:

- Provides equidistant antennas.
- Distance from center to vertex equals length of side.

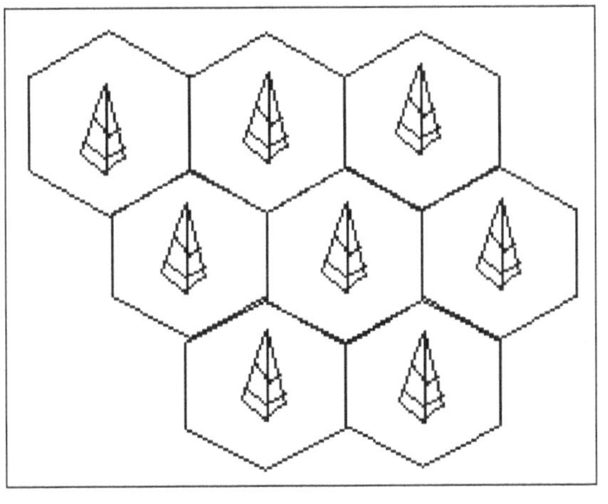

Frequency Reuse

Frequency reusing is the concept of using the same radio frequencies within a given area that are separated by considerable distance, with minimal interference, to establish communication. Frequency reuse offers the following benefits:

- Allows communications within cell on a given frequency.
- Limits escaping power to adjacent cells.
- Allows re-use of frequencies in nearby cells.
- Uses same frequency for multiple conversations.
- 10 to 50 frequencies per cell.

For example, when N cells are using the same number of frequencies and K be the total number of frequencies used in systems. Then each cell frequency is calculated by using the formulae K/N.

In Advanced Mobile Phone Services (AMPS) when K = 395 and N = 7, then frequencies per cell on an average will be 395/7 = 56. Here, cell frequency is 56.

Global Area Network

A global area network (GAN) refers to a network composed of different interconnected networks that cover an unlimited geographical area. The term is loosely synonymous with Internet, which is considered a global area network.

Unlike local area networks (LAN) and wide area networks (WAN), GANs cover a large geographical area. Because a GAN is used to support mobile communication across a number of wireless LANs, the key challenge for any GAN is transferring user communications from one local coverage area to the next.

The most sought-after GAN type is a broadband GAN. The broadband GAN is a global satellite Internet network that uses portable terminals for telephony. The terminals connect laptop computers located in remote areas to broadband Internet.

References

- Wireless-lan-introduction: javatpoint.com, Retrieved 30, March 2020
- Types-of-Wireless-LANs, information-technology, technical-references: idc-online.com, Retrieved 05, April 2020
- Wireless-Ad-Hoc-Networks-Types-Applications-Security-Goals-323967539: researchgate.net, Retrieved 30, June 2020
- How-does-a-mobile-network-work-2373338: lifewire.com, Retrieved 10, August 2020
- Wireless-communication-cellular-networks, wireless-communication: tutorialspoint.com, Retrieved 11, May 2020

4

Forms of Wireless Communication

Mobile telephony is the provision of telephone services to phones which can be moved from one place to another. Bluetooth is also a wireless technology which is used to transfer data between two devices over short distances using ultra high frequency radiowaves and building personal area network. All the diverse aspects of wireless communications have been carefully analyzed in this chapter.

Mobile Telephony

A mobile phone is an electronic device used for mobile telecommunications over a cellular network of specialized base stations known as cell sites. A cell phone offers full Duplex Communication and transfer the link when the user moves from one cell to another. As the phone user moves from one cell area to another, the system automatically commands the mobile phone and a cell site with a stronger signal, to switch on to a new frequency in order to keep the link.

Mobile phone is primarily designed for Voice communication. In addition to the standard voice function, new generation mobile phones support many additional services, and accessories, such as SMS for text messaging, email, packet switching for access to the Internet, gaming, Bluetooth, camera with video recorder and MMS for sending and receiving photos and video, MP3 player, radio and GPS.

The different types of mobile communication systems are mobile two-way radio, public land radio, mobile telephone and amateur (HAM) radio.

Mobile two-way radios are one-to-many communication systems that operate in half-duplex mode, i.e., push to talk. The most common among this type is citizen band (CB) radio, which uses amplitude modulation (AM). It operates in the frequency range of 26-27.1 MHz having 40 channels of 10 kHz. It is a non-commercial service that uses a press-to-talk switch. It can be amplitude modulated having double-sideband suppressed carrier or single-sideband suppressed carrier.

Public land mobile radio is a twoway FM radio system, used in police, fireand municipal agencies. It is limited to small geographical areas.

Amateur (HAM) radios cover a broad frequency band from 1.8 MHz to above 30 MHz. These include continuous wave (CW), AM, FM, radio teleprinter, HF slow-scan still picture TV, VHF or UHF slow-scan or fast-scan TV, facsimile, frequency-shift keying and amplitude-shift keying.

Mobile telephones offer full-duplex transmission. These are one-to-one systems that permit two simultaneous transmissions. For privacy, each mobile unit carries a unique telephone number.

Amateur (HAM) radios cover a broad frequency band from 1.8 MHz to above 30 MHz. These include continuous wave (CW), AM, FM, radio teleprinter, HF slow-scan still picture TV, VHF or UHF slow-scan or fast-scan TV, facsimile, frequency-shift keying and amplitude-shift keying.

Signal Frequency in Cell Phone

The cellular system is the division of an area into small cells. This allows extensive frequency reuse across that area, so that many people can use cell phones simultaneously. Cellular networks has a number of advantages like increased capacity, reduced power usage, larger coverage area, reduced interference from other signals etc.

FDMA and CDMA Systems

Frequency Division Multiple Access (FDMA) and Code Division Multiple Access (CDMA) were developed to distinguish signals from several different transmitters. In FDMA, the transmitting and receiving frequencies used in each cell are different from the frequencies used in the neighboring cells. The principle of CDMA is more complex and the distributed transceivers can select one cell and listen to it. Other methods include Polarization Division Multiple Access (PDMA) and Time Division Multiple Access (TDMA). Time division multiple access is used in combination with either FDMA or CDMA to give multiple channels within the coverage area of a single cell.

Cellular Systems

Mobile phones receive and send radio signals with any number of cell site base stations fitted with microwave antennas. These sites are usually mounted on a tower, pole or building, located throughout populated areas, then connected to a cabled communication network and switching system. The phones have a low-power transceiver that transmits voice and data to the nearest cell sites, normally not more than 8 to 13 km (approximately 5 to 8 miles) away. In areas of low coverage, a cellular repeater may be used, which uses a long distance high-gain dish antenna or yagi antenna to communicate with a cell tower far outside of normal range, and a repeater to rebroadcast on a small short-range local antenna that allows any cellphone within a few meters to function properly.

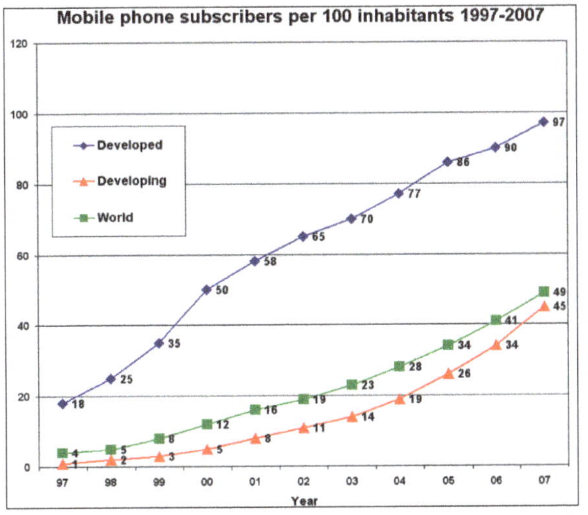

Mobile phone subscriptions, not subscribers, per 100 inhabitants.

When the mobile phone or data device is turned on, it registers with the mobile telephone exchange, or switch, with its unique identifiers, and can then be alerted by the mobile switch when there is an incoming telephone call. The handset constantly listens for the strongest signal being received from the surrounding base stations, and is able to switch seamlessly between sites. As the user moves around the network, the "handoffs" are performed to allow the device to switch sites without interrupting the call.

Cell sites have relatively low-power (often only one or two watts) radio transmitters which broadcast their presence and relay communications between the mobile handsets and the switch. The switch in turn connects the call to another subscriber of the same wireless service provider or to the public telephone network, which includes the networks of other wireless carriers. Many of these sites are camouflaged to blend with existing environments, particularly in scenic areas.

The dialogue between the handset and the cell site is a stream of digital data that includes digitised audio (except for the first generation analog networks). The technology that achieves this depends on the system which the mobile phone operator has adopted. The technologies are grouped by generation. The first-generation systems started in 1979 with Japan, are all analog and include AMPS and NMT. Second-generation systems, started in 1991 in Finland, are all digital and include GSM, CDMA and TDMA.

The nature of cellular technology renders many phones vulnerable to 'cloning': anytime a cell phone moves out of coverage (for example, in a road tunnel), when the signal is re-established, the phone sends out a 're-connect' signal to the nearest cell-tower, identifying itself and signalling that it is again ready to transmit. With the proper equipment, it's possible to intercept the re-connect signal and encode the data it contains into a 'blank' phone—in all respects, the 'blank' is then an exact duplicate of the real phone and any calls made on the 'clone' will be charged to the original account. This problem was widespread with the first generation analogue technology, however the modern digital standards such as GSM greatly improve security and make cloning hard to achieve.

In an effort to limit the potential harm from having a transmitter close to the user's body, the first fixed/mobile cellular phones that had a separate transmitter, vehicle-mounted antenna, and handset (known as *car phones* and *bag phones*) were limited to a maximum 3 watts Effective Radiated Power. Modern *handheld* cellphones which must have the transmission antenna held inches from the user's skull are limited to a maximum transmission power of 0.6 watts ERP. Regardless of the potential biological effects, the reduced transmission range of modern handheld phones limits their usefulness in rural locations as compared to car/bag phones, and handhelds require that cell towers are spaced much closer together to compensate for their lack of transmission power.

Usage

By Civilians

This Railfone found on some Amtrak trains in North America uses cellular technology.

An increasing number of countries, particularly in Europe, now have more mobile phones than people. According to the figures from Eurostat, the European Union's in-house statistical office, Luxembourg had the highest mobile phone penetration rate at 158 mobile subscriptions per 100 people, closely followed by Lithuania and Italy. In Hong Kong the penetration rate reached 139.8% of the population in July 2007. Over 50 countries have mobile phone subscription penetration rates higher than that of the population and the Western European average penetration rate was 110% in 2007 (source Informa 2007). Canada currently has the lowest rates of mobile phone penetrations in the industrialised world at 58%.

There are over five hundred million active mobile phone accounts in China, as of 2007, but the total penetration rate there still stands below 50%. The total number of mobile phone subscribers in the world was estimated at 2.14 billion in 2005. The subscriber count reached 2.7 billion by end of 2006 according to Informa, and 3.3 billion by

November, 2007, thus reaching an equivalent of over half the planet's population. Around 80% of the world's population has access to mobile phone coverage, as of 2006. This figure is expected to increase to 90% by the year 2010.

In some developing countries with little "landline" telephone infrastructure, mobile phone use has quadrupled in the last decade. The rise of mobile phone technology in developing countries is often cited as an example of the leapfrog effect. Many remote regions in the third world went from having no telecommunications infrastructure to having satellite based communications systems. At present, Africa has the largest growth rate of cellular subscribers in the world, its markets expanding nearly twice as fast as Asian markets. The availability of prepaid or 'pay-as-you-go' services, where the subscriber is not committed to a long term contract, has helped fuel this growth in Africa as well as in other continents.

On a numerical basis, India is the largest growth market, adding about 6 million mobile phones every month. It currently has a mobile subscriber base of 937.06 million mobile phones.

Traffic

Since the world is operating quickly to 3G and 4G networks, mobile traffic through video is heading high. It is expected that by end of 2018, the global traffic will reach an annual rate of 190 exabytes/year. This is the result of people shifting to smart phones now-a-days. It is predicted by 2018, mobile traffic will reach by 10 billion connections with 94% traffic comes from Smartphones, laptops and tablets. Also 69% of mobile traffic from Videos since we have high definition screens available in smart phones and 176.9 wearable devices to be at use. Apparently, 4G will be dominating the traffic by 51% of total mobile data by 2018.

By Government Agencies

Law Enforcement

Law enforcement have used mobile phone evidence in a number of different ways. Evidence about the physical location of an individual at a given time can be obtained by triangulating the individual's cellphone between several cellphone towers. This triangulation technique can be used to show that an individual's cellphone was at a certain location at a certain time. The concerns over terrorism and terrorist use of technology prompted an inquiry by the British House of Commons Home Affairs Select Committee into the use of evidence from mobile phone devices, prompting leading mobile telephone forensic specialists to identify forensic techniques available in this area. NIST have published guidelines and procedures for the preservation, acquisition, examination, analysis, and reporting of digital information present on mobile phones can be found under the NIST Publication SP800-101.

In the UK in 2000 it was claimed that recordings of mobile phone conversations made on the day of the Omagh bombing were crucial to the police investigation. In particular, calls made on two mobile phones which were tracked from south of the Irish border to Omagh and back on the day of the bombing, were considered of vital importance.

Further example of criminal investigations using mobile phones is the initial location and ultimate identification of the terrorists of the 2004 Madrid train bombings. In the attacks, mobile phones had been used to detonate the bombs. However, one of the bombs failed to detonate, and the SIM card in the corresponding mobile phone gave the first serious lead about the terrorists to investigators. By tracking the whereabouts of the SIM card and correlating other mobile phones that had been registered in those areas, police were able to locate the terrorists.

Disaster Response

The Finnish government decided in 2005 that the fastest way to warn citizens of disasters was the mobile phone network. In Japan, mobile phone companies provide immediate notification of earthquakes and other natural disasters to their customers free of charge. In the event of an emergency, disaster response crews can locate trapped or injured people using the signals from their mobile phones. An interactive menu accessible through the phone's Internet browser notifies the company if the user is safe or in distress. In Finland rescue services suggest hikers carry mobile phones in case of emergency even when deep in the forests beyond cellular coverage, as the radio signal of a cellphone attempting to connect to a base station can be detected by overflying rescue aircraft with special detection gear. Also, users in the United States can sign up through their provider for free text messages when an AMBER Alert goes out for a missing person in their area.

However, most mobile phone networks operate close to capacity during normal times, and spikes in call volumes caused by widespread emergencies often overload the system just when it is needed the most. Examples reported in the media where this has occurred include the September 11, 2001 attacks, the 2003 Northeast blackouts, the 2005 London Tube bombings, Hurricane Katrina, the 2006 Kiholo Bay earthquake, and the 2007 Minnesota bridge collapse.

Under FCC regulations, all mobile telephones must be capable of dialing emergency telephone numbers, regardless of the presence of a SIM card or the payment status of the account.

Impact on Society

Human Health

Since the introduction of mobile phones, concerns (both scientific and public) have been raised about the potential health impacts from regular use. But by 2008, American mobile phones transmitted and received more text messages than phone calls.

Numerous studies have reported no significant relationship between mobile phone use and health, but the effect of mobile phone usage on health continues to be an area of public concern.

For example, at the request of some of their customers, Verizon created usage controls that meter service and can switch phones off, so that children could get some sleep. There have also been attempts to limit use by persons operating moving trains or automobiles, coaches when writing to potential players on their teams, and movie theater audiences. By one measure, nearly 40% of automobile drivers aged 16 to 30 years old text while driving, and by another, 40% of teenagers said they could text blindfolded.

18 studies have been conducted on the link between cell phones and brain cancer; A review of these studies found that cell phone use of 10 years or more "give a consistent pattern of an increased risk for acoustic neuroma and glioma". The tumors are found mostly on the side of the head that the mobile phone is in contact with. In July 2008, Dr. Ronald Herberman, director of the University of Pittsburgh Cancer Institute, warned about the radiation from mobile phones. He stated that there was no definitive proof of the link between mobile phones and brain tumors but there was enough studies that mobile phone usage should be reduced as a precaution. To reduce the amount of radiation being absorbed hands free devices can be used or texting could supplement calls. Calls could also be shortened or limit mobile phone usage in rural areas. Radiation is found to be higher in areas that are located away from mobile phone towers.

According to Reuters, The British Association of Dermatologists is warning of a rash occurring on people's ears or cheeks caused by an allergic reaction from the nickel surface commonly found on mobile devices' exteriors. There is also a theory it could even occur on the fingers if someone spends a lot of time text messaging on metal menu buttons. In 2008, Lionel Bercovitch of Brown University in Providence, Rhode Island, and his colleagues tested 22 popular handsets from eight different manufacturers and found nickel on 10 of the devices.

Human Behaviour

Culture and Customs

Cellular phones allow people to communicate from almost anywhere at their leisure.

Between the 1980s and the 2000s, the mobile phone has gone from being an expensive item used by the business elite to a pervasive, personal communications tool for the general population. In most countries, mobile phones outnumber land-line phones, with fixed landlines numbering 1.3 billion but mobile subscriptions 3.3 billion at the end of 2007.

In many markets from Japan and South Korea, to Europe, to Malaysia, Singapore, Taiwan and Hong Kong, most children age 8-9 have mobile phones and the new accounts are now opened for customers aged 6 and 7. Where mostly parents tend to give hand-me-down used phones to their youngest children, in Japan already new cameraphones are on the market whose target age group is under 10 years of age, introduced by KDDI in February 2007. The USA also lags on this measure, as in the US so far, about half of all children have mobile phones. In many young adults' households it has supplanted the land-line phone. Mobile phone usage is banned in some countries, such as North Korea and restricted in some other countries such as Burma.

Given the high levels of societal mobile phone service penetration, it is a key means for people to communicate with each other. The SMS feature spawned the "texting" sub-culture amongst younger users. In December 1993, the first person-to-person SMS text message was transmitted in Finland. Currently, texting is the most widely used data service; 1.8 billion users generated $80 billion of revenue in 2006 (source ITU). Many phones offer Instant Messenger services for simple, easy texting. Mobile phones have Internet service (e.g. NTT DoCoMo's i-mode), offering text messaging via e-mail in Japan, South Korea, China, and India. Most mobile internet access is much different from computer access, featuring alerts, weather data, e-mail, search engines, instant messages, and game and music downloading; most mobile internet access is hurried and short.

Because mobile phones are often used publicly, social norms have been shown to play a major role in the usage of mobile phones. Furthermore, the mobile phone can be a fashion totem custom-decorated to reflect the owner's personality and may be a part of their self-identity. This aspect of the mobile telephony business is, in itself, an industry, e.g. ringtone sales amounted to $3.5 billion in 2005. Mobile phone use on aircraft is starting to be allowed with several airlines already offering the ability to use phones during flights. Mobile phone use during flights used to be prohibited and many airlines still claim in their in-plane announcements that this prohibition is due to possible interference with aircraft radio communications. Shut-off mobile phones do not interfere with aircraft avionics. The recommendation why phones should not be used during take-off and landing, even on planes that allow calls or messaging, is so that passengers pay attention to the crew for any possible accident situations, as most aircraft accidents happen on take-off and landing.

Etiquette

Mobile phone use can be an important matter of social discourtesy: phones ringing during funerals or weddings; in toilets, cinemas and theatres. Some book shops, libraries, bathrooms, cinemas, doctors' offices and places of worship prohibit their use, so that other patrons will not be disturbed by conversations. Some facilities install signal-jamming equipment to prevent their use, although in many countries, including the US, such equipment is illegal.

Many US cities with subway transit systems underground are studying or have implemented mobile phone reception in their underground tunnels for their riders, and trains, particularly those involving long-distance services, often offer a "quiet carriage" where phone use is prohibited, much like the designated non-smoking carriage of the past. Most schools in the United States and Europe and Canada have prohibited mobile phones in the classroom, or in school in an effort to limit class disruptions.

A working group made up of Finnish telephone companies, public transport operators and communications authorities has launched a campaign to remind mobile phone users of courtesy, especially when using mass transit—what to talk about on the phone, and how to. In particular, the campaign wants to impact loud mobile phone usage as well as calls regarding sensitive matters.

Use by Drivers

The use of mobile phones by people who are driving has become increasingly common, for example as part of their job, as in the case of delivery drivers who are calling a client, or socially as for commuters who are chatting with a friend. While many drivers have embraced the convenience of using their cellphone while driving, some jurisdictions have made the practice against the law, such as Australia, the Canadian provinces of British Columbia, Quebec, Ontario, Nova Scotia, and Newfoundland and Labrador as well as the United Kingdom, consisting of a zero-tolerance system operated in Scotland and a warning system operated in England, Wales, and Northern Ireland. Officials from these jurisdictions argue that using a mobile phone while driving is an impediment to vehicle operation that can increase the risk of road traffic accidents.

Studies have found vastly different relative risks (RR). Two separate studies using case-crossover analysis each calculated RR at 4, while an epidemiological cohort study found RR, when adjusted for crash-risk exposure, of 1.11 for men and 1.21 for women.

A simulation study from the University of Utah Professor David Strayer compared drivers with a blood alcohol content of 0.08% to those conversing on a cell phone, and after controlling for driving difficulty and time on task, the study concluded that cell phone drivers exhibited greater impairment than intoxicated drivers. Meta-analysis by The Canadian Automobile Association and The University of Illinois found that response time while using both hands-free and hand-held phones was approximately 0.5 standard deviations higher than normal driving (i.e., an average driver, while talking on a cell phone, has response times of a driver in roughly the 40th percentile).

Driving while using a hands-free device is not safer than driving while using a hand-held phone, as concluded by case-crossover studies. epidemiological studies, simulation studies, and meta-analysis. Even with this information, California initiated new Wireless Communications Device Law (effective January 1, 2009) makes it an infraction to write, send, or read text-based communication on an electronic wireless communications device, such as a cell phone, while driving a motor vehicle. Two additional laws dealing with the use of wireless telephones while driving went into effect July 1, 2008. The first law prohibits all drivers from using a handheld wireless telephone while operating a motor vehicle. The law allows a driver to use a wireless telephone to make emergency calls to a law enforcement agency, a medical provider, the fire department, or other emergency services agency. The base fine for the FIRST offense is $20 and $50 for subsequent convictions. With penalty assessments, the fine can be more than triple the base fine amount. videos about California cellular phone laws; with captions (California Vehicle Code [VC] §23123). Motorists 18 and over may use a "hands-free device. The second law effective July 1, 2008, prohibits drivers under the age of 18 from using a wireless telephone or hands-free device while operating a motor vehicle (VC §23124)The consistency of increased crash risk between hands-free and hand-held phone use is at odds with legislation in over 30 countries that prohibit hand-held phone use but allow hands-free. Scientific literature is mixed on the dangers of talking on a phone versus those of talking with a passenger, with the Accident Research Unit at the University of Nottingham finding that the number of utterances was usually higher for mobile calls when compared to blindfolded and non-blindfolded passengers, but the University of Illinois meta-analysis concluding that passenger conversations were just as costly to driving performance as cell phone ones.

Use on Aircraft

As of 2007, several airlines are experimenting with base station and antenna systems installed on the airplane, allowing low power, short-range connection of any phones aboard to remain connected to the aircraft's base station. Thus, they would not attempt connection to the ground base stations as during take off and landing. Simultaneously, airlines may offer phone services to their travelling passengers either as full voice and data services, or initially only as SMS text messaging and similar services. The Australian airline Qantas is the first airline to run a test aeroplane in this configuration in the autumn of 2007. Emirates has announced plans to allow limited mobile phone usage on some flights. However, in the past, commercial airlines have prevented the use of cell phones and laptops, due to the assertion that the frequencies emitted from these devices may disturb the radio waves contact of the airplane.

On March 20, 2008, an Emirates flight was the first time voice calls have been allowed in-flight on commercial airline flights. The breakthrough came after the European Aviation Safety Agency (EASA) and the United Arab Emirates-based General Civil Aviation Authority (GCAA) granted full approval for the AeroMobile system to be used on Emirates. Passengers were able to make and receive voice calls as well as use text messaging. The system automatically came into operation as the Airbus A340-300 reached cruise altitude. Passengers wanting to use the service received a text message welcoming them to the AeroMobile system when they first switched their phones on. The approval by EASA has established that GSM phones are safe to use on airplanes, as the AeroMobile system does not require the modification of aircraft components deemed "sensitive," nor does it require the use of modified phones.

In any case, there are inconsistencies between practices allowed by different airlines and even on the same airline in different countries. For example, Delta Air Lines may allow the use of mobile phones immediately after landing on a domestic flight within the US, whereas they may state "not until the doors are open" on an international flight arriving in the Netherlands. In April 2007 the US Federal Communications Commission officially prohibited passengers' use of cell phones during a flight.

In a similar vein, signs are put up in many countries, such as Canada, the UK and the U.S., at petrol stations prohibiting the use of mobile phones, due to possible safety issues. However, it is unlikely that mobile phone use can cause any problems, and in fact "petrol station employees have themselves spread the rumour about alleged incidents."

Environmental Impacts

Cellular antenna disguised to look like a tree.

Like all high structures, cellular antenna masts pose a hazard to low flying aircraft. Towers over a certain height or towers that are close to airports or heliports are normally required to have warning lights. There have been reports that warning lights on cellular masts, TV-towers and other high structures can attract and confuse birds. US authorities estimate that millions of birds are killed near communication towers in the country each year.

Some cellular antenna towers have been camouflaged to make them less obvious on the horizon, and make them look more like a tree.

An example of the way mobile phones and mobile networks have sometimes been perceived as a threat is the widely reported and later discredited claim that mobile phone masts are associated with the Colony Collapse Disorder (CCD) which has reduced bee hive numbers by up to 75% in many areas, especially near cities in the US. The Independent newspaper cited a scientific study claiming it provided evidence for the theory that mobile phone masts *are* a major cause in the collapse of bee populations, with controlled experiments demonstrating a rapid and catastrophic effect on individual hives near masts. Mobile phones were in fact not covered in the study, and the original researchers have since emphatically disavowed any connection between their research, mobile phones, and CCD, specifically indicating that the Independent article had misinterpreted their results and created "a horror story". While the initial claim of damage to bees was widely reported, the corrections to the story were almost non-existent in the media.

There are more than 500 million used mobile phones in the US sitting on shelves or in landfills, and it is estimated that over 125 million will be discarded this year alone. The problem is growing at a rate of more than two million phones per week, putting tons of toxic waste into landfills daily. Several companies offer to buy back and recycle mobile phones from users. In the United States many unwanted but working mobile phones are donated to women's shelters to allow emergency communication.

Tariff Models

Mobile phone shop in Uganda.

Payment Methods

There are two principal ways to pay for mobile telephony: the 'pay-as-you-go' model where conversation time is purchased and added to a phone unit via an Internet account or in shops or ATMs, or the contract model where bills are paid by regular intervals after the service has been consumed. It is increasingly common for a consumer to purchase a basic package and then bolt-on services and functionality to create a subscription customised to the users needs.

Pay as you go (also known as "pre-pay" or "prepaid") accounts were invented simultaneously in Portugal and Italy and today form more than half of all mobile phone subscriptions. USA, Canada, Costa Rica, Japan, Israel and Finland are among the rare countries left where most phones are still contract-based.

Incoming Call Charges

In the early days of mobile telephony, the operators (carriers) charged for all air time consumed by the mobile phone user, which included both outbound and inbound telephone calls. As mobile phone adoption rates increased, competition between operators meant that some decided not to charge for incoming calls in some markets (also called "calling party pays").

The European market adopted a calling party pays model throughout the GSM environment and soon various other GSM markets also started to emulate this model.

In Hong Kong, Singapore, Canada, and the United States, it is common for the party receiving the call to be charged per minute, although a few carriers are beginning to offer unlimited received phone calls. This is called the "Receiving Party Pays" model. In China, it was reported that both of its two operators will adopt the caller-pays approach as early as January 2007.

One disadvantage of the receiving party pays systems is that phone owners keep their phones turned off to avoid receiving unwanted calls, which results in the total voice usage rates (and profits) in Calling Party Pays countries outperform those in Receiving Party Pays countries. To avoid the problem of users keeping their phone turned off, most Receiving Party Pays countries have either switched to Calling Party Pays, or their carriers offer additional incentives such as a large number of monthly minutes at a sufficiently discounted rate to compensate for the inconvenience.

Note that when a user roaming in another country, international roaming tariffs apply to all calls received, regardless of the model adopted in the home country.

Bluetooth

A Bluetooth technology is a high speed low powered wireless technology link that is designed to connect phones or other portable equipment together. It is a specification (IEEE 802.15.1) for the use of low power radio communications to link phones, computers and other network devices over short distance without wires. Wireless signals transmitted with Bluetooth cover short distances, typically up to 30 feet (10 meters).

It is achieved by embedded low cost transceivers into the devices. It supports on the frequency band of 2.45GHz and can support upto 721KBps along with three voice channels. This frequency band has been set aside by international agreement for the use of industrial, scientific and medical devices (ISM).rd-compatible with 1.0 devices.

Bluetooth can connect up to "eight devices" simultaneously and each device offers a unique 48 bit address from the IEEE 802 standard with the connections being made point to point or multipoint.

Now a given feature on everything from smartphones to in-vehicle entertainment systems, Bluetooth has an interesting history and working, proving how versatile this communication standard is. It is managed by the Bluetooth Special Interest Group (SIG), a non-profit, non stock company which mainly deals with setting standards, licensing and advancing Bluetooth capabilities. Before this it was standardized by the IEEE as IEEE 802.15.1, but since the SIG, that convention is not used.

Originally conceived as an alternative to the popular RS-232 cables, Bluetooth was invented by telecom giant Ericsson in 1994. Supported by Intel, it then created the above mentioned body called the Bluetooth Special Interest Group that oversaw all development and licensing. For any company to use the standard or market products with the technology, they had to become a member of the SIG. As a result, the SIG is today over 20,000 members strong! However, not all members have the same power, there is a division called 'Promoter Members' for the 'elite members', so to speak. Members from this group have a say in the direction the company is going, and have generally had a hand in the development of the standard as a whole. The Promoter members are:

- Ericsson
- Intel
- Microsoft
- Nokia
- Lenovo
- Toshiba
- Motorola

The Bluetooth Special Interest Group has a set of terms and conditions that must be followed by all members, and also has a set of compliance guidelines for all devices.

Now, to the very interesting name. Ericsson, the company that started Bluetooth is from Sweden, which is part of the Scandinavian region, a historical and cultural-linguistic part of Europe. The name comes from an epithet of a tenth-century King of Denmark and Norway named Harald "Bluetooth" Gormsson. In the local language, he was called *Blåtand* or *Blåtann,* which translated in English became 'Bluetooth'. He was known for uniting the Vikings in ages past, from which the idea of the communication standard came, something that was a single unifying standard for mobile technologies. The logo, in fact, is a combination of two Nordic runes called 'Hagall' and 'Bjarkan', which were the initials of King Harald "Bluetooth" Gormsson.

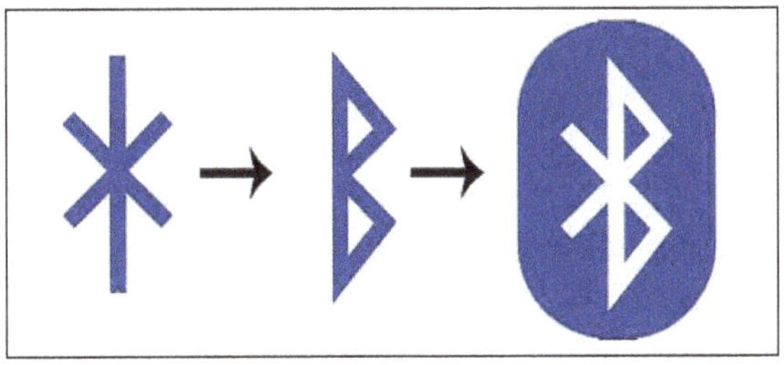

Implementation

Bluetooth operates at frequencies between 2402 and 2480 MHz, or 2400 and 2483.5 MHz including guard bands 2 MHz wide at the bottom end and 3.5 MHz wide at the top. This is in the globally unlicensed (but not unregulated) industrial, scientific and medical (ISM) 2.4 GHz short-range radio frequency band. Bluetooth uses a radio technology called frequency-hopping spread spectrum. Bluetooth divides transmitted data into packets, and transmits each packet on one of 79 designated Bluetooth channels. Each channel has a bandwidth of 1 MHz. It usually performs 800 hops per second, with Adaptive Frequency-Hopping (AFH) enabled. Bluetooth Low Energy uses 2 MHz spacing, which accommodates 40 channels.

Originally, Gaussian frequency-shift keying (GFSK) modulation was the only modulation scheme available. Since the introduction of Bluetooth 2.0+EDR, π/4-DQPSK (differential quadrature phase shift keying) and 8DPSK modulation may also be used between compatible devices. Devices functioning with GFSK are said to be operating in basic rate (BR) mode where an instantaneous bit rate of 1 Mbit/s is possible. The term Enhanced Data Rate (EDR) is used to describe π/4-DPSK and 8DPSK schemes, each giving 2 and 3 Mbit/s respectively. The combination of these (BR and EDR) modes in Bluetooth radio technology is classified as a "BR/EDR radio".

Bluetooth is a packet-based protocol with a master/slave architecture. One master may communicate with up to seven slaves in a piconet. All devices share the master's clock. Packet exchange is based on the basic clock, defined by the master, which ticks at 312.5 µs intervals. Two clock ticks make up a slot of 625 µs, and two slots make up a slot pair of 1250 µs. In the simple case of single-slot packets the master transmits in even slots and receives in odd slots. The slave, conversely, receives in even slots and transmits in odd slots. Packets may be 1, 3 or 5 slots long, but in all cases the master's transmission begins in even slots and the slave's in odd slots.

The above is valid for "classic" BT. Bluetooth Low Energy, introduced in the 4.0 specification, uses the same spectrum but somewhat differently.

Communication and Connection

A master BR/EDR Bluetooth device can communicate with a maximum of seven devices in a piconet (an ad-hoc computer network using Bluetooth technology), though not all devices reach this maximum. The devices can switch roles, by agreement, and the slave can become the master (for example, a headset initiating a connection to a phone necessarily begins as master—as initiator of the connection—but may subsequently operate as slave).

The Bluetooth Core Specification provides for the connection of two or more piconets to form a scatternet, in which certain devices simultaneously play the master role in one piconet and the slave role in another.

At any given time, data can be transferred between the master and one other device (except for the little-used broadcast mode). The master chooses which slave device to address; typically, it switches rapidly from one device to another in a round-robin fashion. Since it is the master that chooses which slave to address, whereas a slave is (in theory) supposed to listen in each receive slot, being a master is a lighter burden than being a slave. Being a master of seven slaves is possible; being a slave of more than one master is possible. The specification is vague as to required behavior in scatternets.

Uses

Ranges of Bluetooth devices by class			
Class	Max. permitted power		Typ. range (m)
	(mW)	(dBm)	
1	100	20	~100
2	2.5	4	~10
3	1	0	~1
4	0.5	−3	~0.5

Bluetooth is a standard wire-replacement communications protocol primarily designed for low-power consumption, with a short range based on low-cost transceiver microchips in each device. Because the devices use a radio (broadcast) communications system, they do not have to be in visual line of sight of each other; however, a *quasi optical* wireless path must be viable. Range is power-class-dependent, but effective ranges vary in practice.

Officially Class 3 radios have a range of up to 1 metre (3 ft), Class 2, most commonly found in mobile devices, 10 metres (33 ft), and Class 1, primarily for industrial use cases,100 metres (300 ft). Bluetooth Marketing qualifies that Class 1 range is in most cases 20–30 metres (66–98 ft), and Class 2 range 5–10 metres (16–33 ft). The actual range achieved by a given link will depend on the qualities of the devices at both ends of the link, as well as the air conditions in between, and other factors.

The effective range varies depending on propagation conditions, material coverage, production sample variations, antenna configurations and battery conditions. Most Bluetooth applications are for indoor conditions, where attenuation of walls and signal fading due to signal reflections make the range far lower than specified line-of-sight ranges of the Bluetooth products.

Most Bluetooth applications are battery-powered Class 2 devices, with little difference in range whether the other end of the link is a Class 1 or Class 2 device as the lower-powered device tends to set the range limit. In some cases the effective range of the data link can be extended when a Class 2 device is connecting to a Class 1 transceiver with both higher sensitivity and transmission power than a typical Class 2 device. Mostly, however, the Class 1 devices have a similar sensitivity to Class 2 devices. Connecting two Class 1 devices with both high sensitivity and high power can allow ranges far in excess of the typical 100m, depending on the throughput required by the application. Some such devices allow open field ranges of up to 1 km and beyond between two similar devices without exceeding legal emission limits.

The Bluetooth Core Specification mandates a range of not less than 10 metres (33 ft), but there is no upper limit on actual range. Manufacturers' implementations can be tuned to provide the range needed for each case.

Bluetooth Profile

To use Bluetooth wireless technology, a device must be able to interpret certain Bluetooth profiles, which are definitions of possible applications and specify general behaviors that Bluetooth-enabled devices use to communicate with other Bluetooth devices. These profiles include settings to parameterize and to control the communication from the start. Adherence to profiles saves the time for transmitting the parameters anew before the bi-directional link becomes effective. There are a wide range of Bluetooth profiles that describe many different types of applications or use cases for devices.

List of Applications

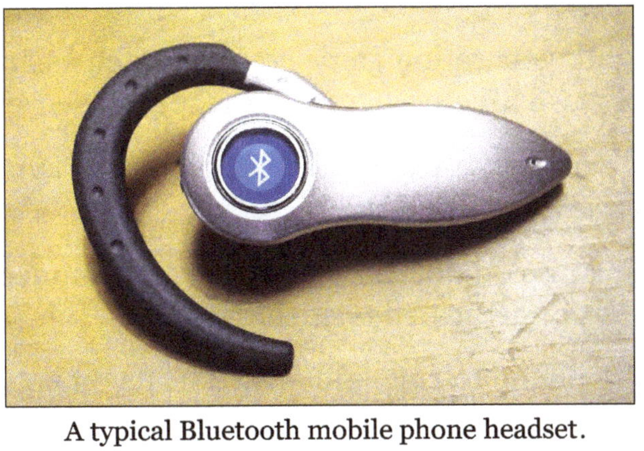

A typical Bluetooth mobile phone headset.

- Wireless control of and communication between a mobile phone and a hands-free headset. This was one of the earliest applications to become popular.
- Wireless control of and communication between a mobile phone and a Bluetooth compatible car stereo system.
- Wireless control of and communication with iOS and Android device phones, tablets and portable wireless speakers.
- Wireless Bluetooth headset and Intercom. Idiomatically, a headset is sometimes called “a Bluetooth”.
- Wireless streaming of audio to headphones with or without communication capabilities.
- Wireless streaming of data collected by Bluetooth-enabled fitness devices to phone or PC.
- Wireless networking between PCs in a confined space and where little bandwidth is required.
- Wireless communication with PC input and output devices, the most common being the mouse, keyboard and printer.
- Transfer of files, contact details, calendar appointments, and reminders between devices with OBEX.
- Replacement of previous wired RS-232 serial communications in test equipment, GPS receivers, medical equipment, bar code scanners, and traffic control devices.
- For controls where infrared was often used.
- For low bandwidth applications where higher USB bandwidth is not required and cable-free connection desired.
- Sending small advertisements from Bluetooth-enabled advertising hoardings to other, discoverable, Bluetooth devices.
- Wireless bridge between two Industrial Ethernet (e.g., PROFINET) networks.
- Seventh and eighth generation game consoles such as Nintendo’s Wii, and Sony’s PlayStation 3 use Bluetooth for their respective wireless controllers.
- Dial-up internet access on personal computers or PDAs using a data-capable mobile phone as a wireless modem.
- Short-range transmission of health sensor data from medical devices to mobile phone, set-top box or dedicated telehealth devices.
- Allowing a DECT phone to ring and answer calls on behalf of a nearby mobile phone.

- Real-time location systems (RTLS) are used to track and identify the location of objects in real time using "Nodes" or "tags" attached to, or embedded in, the objects tracked, and "Readers" that receive and process the wireless signals from these tags to determine their locations.
- Personal security application on mobile phones for prevention of theft or loss of items. The protected item has a Bluetooth marker (e.g., a tag) that is in constant communication with the phone. If the connection is broken (the marker is out of range of the phone) then an alarm is raised. This can also be used as a man overboard alarm. A product using this technology has been available since 2009.
- Calgary, Alberta, Canada's Roads Traffic division uses data collected from travelers' Bluetooth devices to predict travel times and road congestion for motorists.
- Wireless transmission of audio (a more reliable alternative to FM transmitters).
- Live video streaming to the visual cortical implant device by Nabeel Fattah in Newcastle university 2017.

Bluetooth vs. Wi-Fi (IEEE 802.11)

Bluetooth and Wi-Fi (Wi-Fi is the brand name for products using IEEE 802.11 standards) have some similar applications: setting up networks, printing, or transferring files. Wi-Fi is intended as a replacement for high-speed cabling for general local area network access in work areas or home. This category of applications is sometimes called wireless local area networks (WLAN). Bluetooth was intended for portable equipment and its applications. The category of applications is outlined as the wireless personal area network (WPAN). Bluetooth is a replacement for cabling in a variety of personally carried applications in any setting, and also works for fixed location applications such as smart energy functionality in the home (thermostats, etc.).

Wi-Fi and Bluetooth are to some extent complementary in their applications and usage. Wi-Fi is usually access point-centered, with an asymmetrical client-server connection with all traffic routed through the access point, while Bluetooth is usually symmetrical, between two Bluetooth devices. Bluetooth serves well in simple applications where two devices need to connect with minimal configuration like a button press, as in headsets and remote controls, while Wi-Fi suits better in applications where some degree of client configuration is possible and high speeds are required, especially for network access through an access node. However, Bluetooth access points do exist, and ad-hoc connections are possible with Wi-Fi though not as simply as with Bluetooth. Wi-Fi Direct was recently developed to add a more Bluetooth-like ad-hoc functionality to Wi-Fi.

Devices

A Bluetooth USB dongle with a 100 m range.

Bluetooth exists in many products, such as telephones, speakers, tablets, media players, robotics systems, handheld, laptops, console gaming equipment as well as some high definition headsets, modems, and watches. The technology is useful when transferring information between two or more devices that are near each other in low-bandwidth situations. Bluetooth is commonly used to transfer sound data with telephones (i.e., with a Bluetooth headset) or byte data with hand-held computers (transferring files).

Bluetooth protocols simplify the discovery and setup of services between devices. Bluetooth devices can advertise all of the services they provide. This makes using services easier, because more of the security, network address and permission configuration can be automated than with many other network types.

Computer Requirements

A personal computer that does not have embedded Bluetooth can use a Bluetooth adapter that enables the PC to communicate with Bluetooth devices. While some desktop computers and most recent laptops come with a built-in Bluetooth radio, others require an external adapter, typically in the form of a small USB "dongle."

A typical Bluetooth USB dongle.

An internal notebook Bluetooth card (14×36×4 mm).

Unlike its predecessor, IrDA, which requires a separate adapter for each device, Bluetooth lets multiple devices communicate with a computer over a single adapter.

Operating System Implementation

For Microsoft platforms, Windows XP Service Pack 2 and SP3 releases work natively with Bluetooth v1.1, v2.0 and v2.0+EDR. Previous versions required users to install their Bluetooth adapter's own drivers, which were not directly supported by Microsoft. Microsoft's own Bluetooth dongles (packaged with their Bluetooth computer devices) have no external drivers and thus require at least Windows XP Service Pack 2. Windows Vista RTM/SP1 with the Feature Pack for Wireless or Windows Vista SP2 work with Bluetooth v2.1+EDR. Windows 7 works with Bluetooth v2.1+EDR and Extended Inquiry Response (EIR).

The Windows XP and Windows Vista/Windows 7 Bluetooth stacks support the following Bluetooth profiles natively: PAN, SPP, DUN, HID, HCRP. The Windows XP stack can be replaced by a third party stack that supports more profiles or newer Bluetooth versions. The Windows Vista/Windows 7 Bluetooth stack supports vendor-supplied additional profiles without requiring that the Microsoft stack be replaced.

Apple products have worked with Bluetooth since [[Mac OS X v10.2]], which was released in 2002.

Linux has two popular Bluetooth stacks, BlueZ and Affix. The BlueZ stack is included with most Linux kernels and was originally developed by Qualcomm. The Affix stack was developed by Nokia.

FreeBSD features Bluetooth since its v5.0 release.

NetBSD features Bluetooth since its v4.0 release. Its Bluetooth stack was ported to OpenBSD as well, however OpenBSD later removed it for security reasons.

Specifications and Features

The specifications were formalized by the Bluetooth Special Interest Group (SIG).

The SIG was formally announced on 20 May 1998. Today it has a membership of over 30,000 companies worldwide. It was established by Ericsson, IBM, Intel, Toshiba and Nokia, and later joined by many other companies.

All versions of the Bluetooth standards support downward compatibility. That lets the latest standard cover all older versions.

The Bluetooth Core Specification Working Group (CSWG) produces mainly 4 kinds of specifications:

- The Bluetooth Core Specification, release cycle is typically a few years in between.
- Core Specification Addendum (CSA), release cycle can be as tight as a few times per year.
- Core Specification Supplements (CSS), can be released very quickly.
- Errata (Available with a user account: Errata login).

Bluetooth 1.0 and 1.0B

Versions 1.0 and 1.0B had many problems, and manufacturers had difficulty making their products interoperable. Versions 1.0 and 1.0B also included mandatory Bluetooth hardware device address (BD_ADDR) transmission in the Connecting process (rendering anonymity impossible at the protocol level), which was a major setback for certain services planned for use in Bluetooth environments.

Bluetooth 1.1

- Ratified as IEEE Standard 802.15.1–2002
- Many errors found in the v1.0B specifications were fixed.
- Added possibility of non-encrypted channels.
- Received Signal Strength Indicator (RSSI).

Bluetooth 1.2

Major enhancements include:

- Faster Connection and Discovery.
- *Adaptive frequency-hopping spread spectrum (AFH)*, which improves resistance to radio frequency interference by avoiding the use of crowded frequencies in the hopping sequence.
- Higher transmission speeds in practice than in v1.1, up to 721 kbit/s.

- Extended Synchronous Connections (eSCO), which improve voice quality of audio links by allowing retransmissions of corrupted packets, and may optionally increase audio latency to provide better concurrent data transfer.
- Host Controller Interface (HCI) operation with three-wire UART.
- Ratified as IEEE Standard 802.15.1–2005
- Introduced Flow Control and Retransmission Modes for L2CAP.

Bluetooth 2.0 + EDR

This version of the Bluetooth Core Specification was released in 2004. The main difference is the introduction of an Enhanced Data Rate (EDR) for faster data transfer. The bit rate of EDR is 3 Mbit/s, although the maximum data transfer rate (allowing for inter-packet time and acknowledgements) is 2.1 Mbit/s. EDR uses a combination of GFSK and Phase Shift Keying modulation (PSK) with two variants, π/4-DQPSK and 8DPSK. EDR can provide a lower power consumption through a reduced duty cycle.

The specification is published as *Bluetooth v2.0 + EDR*, which implies that EDR is an optional feature. Aside from EDR, the v2.0 specification contains other minor improvements, and products may claim compliance to "Bluetooth v2.0" without supporting the higher data rate. At least one commercial device states "Bluetooth v2.0 without EDR" on its data sheet.

Bluetooth 2.1 + EDR

Bluetooth Core Specification Version 2.1 + EDR was adopted by the Bluetooth SIG on 26 July 2007.

The headline feature of v2.1 is secure simple pairing (SSP): this improves the pairing experience for Bluetooth devices, while increasing the use and strength of security.

Version 2.1 allows various other improvements, including "Extended inquiry response" (EIR), which provides more information during the inquiry procedure to allow better filtering of devices before connection; and sniff subrating, which reduces the power consumption in low-power mode.

Bluetooth 3.0 + HS

Version 3.0 + HS of the Bluetooth Core Specification was adopted by the Bluetooth SIG on 21 April 2009. Bluetooth v3.0 + HS provides theoretical data transfer speeds of up to 24 Mbit/s, though not over the Bluetooth link itself. Instead, the Bluetooth link is used for negotiation and establishment, and the high data rate traffic is carried over a colocated 802.11 link.

The main new feature is AMP (Alternative MAC/PHY), the addition of 802.11 as a high-speed transport. The high-speed part of the specification is not mandatory, and hence only devices that display the "+HS" logo actually support Bluetooth over 802.11 high-speed data transfer. A Bluetooth v3.0 device without the "+HS" suffix is only required to support features introduced in Core Specification Version 3.0 or earlier Core Specification Addendum 1.

L2CAP Enhanced modes:

> Enhanced Retransmission Mode (ERTM) implements reliable L2CAP channel, while Streaming Mode (SM) implements unreliable channel with no retransmission or flow control. Introduced in Core Specification Addendum 1.

Alternative MAC/PHY:

> Enables the use of alternative MAC and PHYs for transporting Bluetooth profile data. The Bluetooth radio is still used for device discovery, initial connection and profile configuration. However, when large quantities of data must be sent, the high-speed alternative MAC PHY 802.11 (typically associated with Wi-Fi) transports the data. This means that Bluetooth uses proven low power connection models when the system is idle, and the faster radio when it must send large quantities of data. AMP links require enhanced L2CAP modes.

Unicast Connectionless Data:

> Permits sending service data without establishing an explicit L2CAP channel. It is intended for use by applications that require low latency between user action and reconnection/transmission of data. This is only appropriate for small amounts of data.

Enhanced Power Control:

> Updates the power control feature to remove the open loop power control, and also to clarify ambiguities in power control introduced by the new modulation schemes added for EDR. Enhanced power control removes the ambiguities by specifying the behaviour that is expected. The feature also adds closed loop power control, meaning RSSI filtering can start as the response is received. Additionally, a "go straight to maximum power" request has been introduced. This is expected to deal with the headset link loss issue typically observed when a user puts their phone into a pocket on the opposite side to the headset.

Ultra-wideband

The high-speed (AMP) feature of Bluetooth v3.0 was originally intended for UWB, but the WiMedia Alliance, the body responsible for the flavor of UWB intended for

Bluetooth, announced in March 2009 that it was disbanding, and ultimately UWB was omitted from the Core v3.0 specification.

On 16 March 2009, the WiMedia Alliance announced it was entering into technology transfer agreements for the WiMedia Ultra-wideband (UWB) specifications. WiMedia has transferred all current and future specifications, including work on future high-speed and power-optimized implementations, to the Bluetooth Special Interest Group (SIG), Wireless USB Promoter Group and the USB Implementers Forum. After successful completion of the technology transfer, marketing, and related administrative items, the WiMedia Alliance ceased operations.

In October 2009 the Bluetooth Special Interest Group suspended development of UWB as part of the alternative MAC/PHY, Bluetooth v3.0 + HS solution. A small, but significant, number of former WiMedia members had not and would not sign up to the necessary agreements for the IP transfer. The Bluetooth SIG is now in the process of evaluating other options for its longer term roadmap.

Bluetooth 4.0 + LE

The Bluetooth SIG completed the Bluetooth Core Specification version 4.0 (called Bluetooth Smart) and has been adopted as of 30 June 2010. It includes *Classic Bluetooth*, *Bluetooth high speed* and *Bluetooth Low Energy* protocols. Bluetooth high speed is based on Wi-Fi, and Classic Bluetooth consists of legacy Bluetooth protocols.

Bluetooth Low Energy, previously known as Wibree, is a subset of Bluetooth v4.0 with an entirely new protocol stack for rapid build-up of simple links. As an alternative to the Bluetooth standard protocols that were introduced in Bluetooth v1.0 to v3.0, it is aimed at very low power applications running off a coin cell. Chip designs allow for two types of implementation, dual-mode, single-mode and enhanced past versions. The provisional names *Wibree* and *Bluetooth ULP* (Ultra Low Power) were abandoned and the BLE name was used for a while. In late 2011, new logos "Bluetooth Smart Ready" for hosts and "Bluetooth Smart" for sensors were introduced as the general-public face of BLE.

Compared to *Classic Bluetooth*, Bluetooth Low Energy is intended to provide considerably reduced power consumption and cost while maintaining a similar communication range. In terms of lengthening the battery life of Bluetooth devices, BLE represents a significant progression.

- In a single-mode implementation, only the low energy protocol stack is implemented. Dialog Semiconductor, STMicroelectronics, AMICCOM, CSR, Nordic Semiconductor and Texas Instruments have released single mode Bluetooth Low Energy solutions.
- In a dual-mode implementation, Bluetooth Smart functionality is integrated into an existing Classic Bluetooth controller. As of March 2011, the following

semiconductor companies have announced the availability of chips meeting the standard: Qualcomm-Atheros, CSR, Broadcom and Texas Instruments. The compliant architecture shares all of Classic Bluetooth's existing radio and functionality resulting in a negligible cost increase compared to Classic Bluetooth.

Cost-reduced single-mode chips, which enable highly integrated and compact devices, feature a lightweight Link Layer providing ultra-low power idle mode operation, simple device discovery, and reliable point-to-multipoint data transfer with advanced power-save and secure encrypted connections at the lowest possible cost.

General improvements in version 4.0 include the changes necessary to facilitate BLE modes, as well the Generic Attribute Profile (GATT) and Security Manager (SM) services with AES Encryption.

Core Specification Addendum 2 was unveiled in December 2011; it contains improvements to the audio Host Controller Interface and to the High Speed (802.11) Protocol Adaptation Layer.

Core Specification Addendum 3 revision 2 has an adoption date of 24 July 2012.

Core Specification Addendum 4 has an adoption date of 12 February 2013.

Bluetooth 4.1

The Bluetooth SIG announced formal adoption of the Bluetooth v4.1 specification on 4 December 2013. This specification is an incremental software update to Bluetooth Specification v4.0, and not a hardware update. The update incorporates Bluetooth Core Specification Addenda (CSA 1, 2, 3 & 4) and adds new features that improve consumer usability. These include increased co-existence support for LTE, bulk data exchange rates—and aid developer innovation by allowing devices to support multiple roles simultaneously.

New features of this specification include:

- Mobile Wireless Service Coexistence Signaling.
- Train Nudging and Generalized Interlaced Scanning.
- Low Duty Cycle Directed Advertising.
- L2CAP Connection Oriented and Dedicated Channels with Credit Based Flow Control.
- Dual Mode and Topology.
- LE Link Layer Topology.
- 802.11n PAL

- Audio Architecture Updates for Wide Band Speech.
- Fast Data Advertising Interval.
- Limited Discovery Time.

Notice that some features were already available in a Core Specification Addendum (CSA) before the release of v4.1.

Bluetooth 4.2

Released on December 2, 2014, it introduces features for the Internet of Things.

The major areas of improvement are:

- Low Energy Secure Connection with Data Packet Length Extension.
- Link Layer Privacy with Extended Scanner Filter Policies.
- Internet Protocol Support Profile (IPSP) version 6 ready for Bluetooth Smart things to support connected home.

Older Bluetooth hardware may receive 4.2 features such as Data Packet Length Extension and improved privacy via firmware updates.

Bluetooth 5

The Bluetooth SIG presented Bluetooth 5 on 16 June 2016. Its new features are mainly focused on emerging Internet of Things technology. The Samsung Galaxy S8 launched with Bluetooth 5 support in April 2017. In September 2017, the iPhone 8, 8 Plus and iPhone X launched with Bluetooth 5 support as well. Apple also integrated 'Bluetooth 5.0' in their new HomePod offering released on February 9, 2018. Marketing drops the point number; so that it is just "Bluetooth 5" (not 5.0 or LE like Bluetooth 4.0). The change is for the sake of "Simplifying our marketing, communicating user benefits more effectively and making it easier to signal significant technology updates to the market."

Bluetooth 5 provides, for BLE, options that can double the speed (2 Mbit/s burst) at the expense of range, or up to fourfold the range at the expense of data rate, and eightfold the data broadcasting capacity of transmissions, by increasing the packet lengths. The increase in transmissions could be important for Internet of Things devices, where many nodes connect throughout a whole house. Bluetooth 5 adds functionality for connectionless services such as location-relevant navigation of low-energy Bluetooth connections.

The major areas of improvement are:

- Slot Availability Mask (SAM)
- 2 Mbit/s PHY for LE

- LE Long Range
- High Duty Cycle Non-Connectable Advertising
- LE Advertising Extensions
- LE Channel Selection Algorithm #2

Features Added in CSA5 – Integrated in v5.0:

- Higher Output Power

The following features were removed in this version of the specification:

- Park State

Technical Information

Architecture

Software:

Seeking to extend the compatibility of Bluetooth devices, the devices that adhere to the standard use as interface between the host device (laptop, phone, etc.) and the Bluetooth device as such (Bluetooth chip) an interface called HCI (Host Controller Interface).

High-level protocols such as the SDP (Protocol used to find other Bluetooth devices within the communication range, also responsible for detecting the function of devices in range), RFCOMM (Protocol used to emulate serial port connections) and TCS (Telephony control protocol) interact with the baseband controller through the L2CAP Protocol (Logical Link Control and Adaptation Protocol). The L2CAP protocol is responsible for the segmentation and reassembly of the packets.

Hardware:

The hardware that makes up the Bluetooth device is made up of two parts. A radio device, responsible for modulating and transmitting the signal; and a digital controller. The digital controller is composed of a CPU, a digital signal processor (DSP) called Link Controller and interfaces with the host device. The Link Controller is responsible for the processing of the baseband and the management of ARQ and physical layer FEC protocols. In addition, it handles the transfer functions (both asynchronous and synchronous), audio coding and data encryption. The CPU of the device is responsible for attending the instructions related to Bluetooth of the host device, in order to simplify its operation. To do this, the CPU runs a software called Link Manager that has the function of communicating with other devices through the LMP protocol.

Bluetooth Protocol Stack

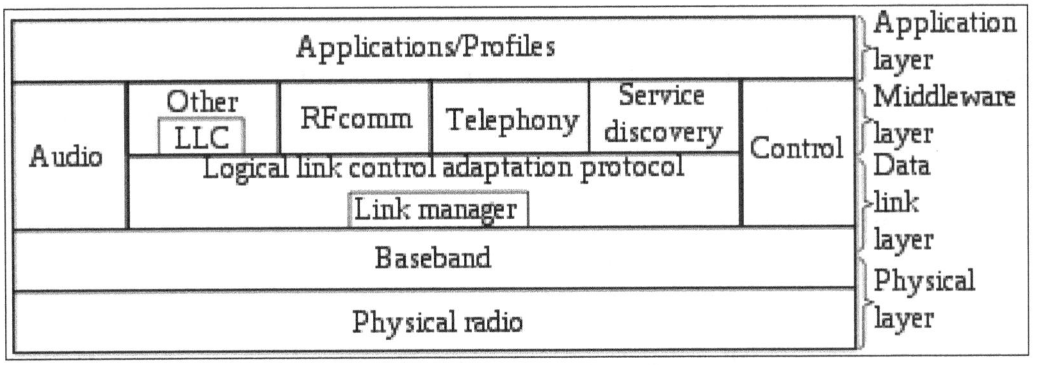

Bluetooth Protocol Stack.

Bluetooth is defined as a layer protocol architecture consisting of core protocols, cable replacement protocols, telephony control protocols, and adopted protocols. Mandatory protocols for all Bluetooth stacks are LMP, L2CAP and SDP. In addition, devices that communicate with Bluetooth almost universally can use these protocols: HCI and RFCOMM.

Link Manager

The Link Manager is the system that manages to establish the connection between devices. It is responsible for the establishment, authentication and configuration of the link. The Link Manager locates other managers and communicates with them thanks to the management protocol of the LMP link. In order to perform its function as a service provider, the LM uses the services included in the link controller (LC, "Link Controller"). The Link Manager Protocol basically consists of a number of PDUs (Protocol Data Units) that are sent from one device to another. The following is a list of supported services: • Transmission and reception of data. • Name request • Request of the link addresses. • Establishment of the connection. • Authentication. • Negotiation of link mode and connection establishment.

Host Controller Interface

The Host Controller Interface provides a command interface for the controller and for the link manager, which allows access to the hardware status and control registers. This interface provides an access layer for all devices Bluetooth The HCI layer of the machine exchanges commands and data with the HCI firmware present in the Bluetooth device. One of the most important HCI tasks that must be performed is the automatic discovery of other Bluetooth devices that are within the coverage radius.

Logical Link Control and Adaptation Protocol

The *Logical Link Control and Adaptation Protocol* (L2CAP) is used to multiplex multiple logical connections between two devices using different higher level protocols. Provides segmentation and reassembly of on-air packets.

In *Basic* mode, L2CAP provides packets with a payload configurable up to 64 kB, with 672 bytes as the default MTU, and 48 bytes as the minimum mandatory supported MTU.

In *Retransmission and Flow Control* modes, L2CAP can be configured either for isochronous data or reliable data per channel by performing retransmissions and CRC checks.

Bluetooth Core Specification Addendum 1 adds two additional L2CAP modes to the core specification. These modes effectively deprecate original Retransmission and Flow Control modes:

- Enhanced Retransmission Mode (ERTM): This mode is an improved version of the original retransmission mode. This mode provides a reliable L2CAP channel.
- Streaming Mode (SM): This is a very simple mode, with no retransmission or flow control. This mode provides an unreliable L2CAP channel.

Reliability in any of these modes is optionally and/or additionally guaranteed by the lower layer Bluetooth BDR/EDR air interface by configuring the number of retransmissions and flush timeout (time after which the radio flushes packets). In-order sequencing is guaranteed by the lower layer.

Only L2CAP channels configured in ERTM or SM may be operated over AMP logical links.

Service Discovery Protocol

The *Service Discovery Protocol* (SDP) allows a device to discover services offered by other devices, and their associated parameters. For example, when you use a mobile phone with a Bluetooth headset, the phone uses SDP to determine which Bluetooth profiles the headset can use (Headset Profile, Hands Free Profile, Advanced Audio Distribution Profile (A2DP) etc.) and the protocol multiplexer settings needed for the phone to connect to the headset using each of them. Each service is identified by a Universally Unique Identifier (UUID), with official services (Bluetooth profiles) assigned a short form UUID (16 bits rather than the full 128).

Radio Frequency Communications

Radio Frequency Communications (RFCOMM) is a cable replacement protocol used for generating a virtual serial data stream. RFCOMM provides for binary data transport and emulates EIA-232 (formerly RS-232) control signals over the Bluetooth baseband layer, i.e. it is a serial port emulation.

RFCOMM provides a simple reliable data stream to the user, similar to TCP. It is used directly by many telephony related profiles as a carrier for AT commands, as well as being a transport layer for OBEX over Bluetooth.

Many Bluetooth applications use RFCOMM because of its widespread support and publicly available API on most operating systems. Additionally, applications that used a serial port to communicate can be quickly ported to use RFCOMM.

Bluetooth Network Encapsulation Protocol

The *Bluetooth Network Encapsulation Protocol* (BNEP) is used for transferring another protocol stack's data via an L2CAP channel. Its main purpose is the transmission of IP packets in the Personal Area Networking Profile. BNEP performs a similar function to SNAP in Wireless LAN.

Audio/Video Control Transport Protocol

The *Audio/Video Control Transport Protocol* (AVCTP) is used by the remote control profile to transfer AV/C commands over an L2CAP channel. The music control buttons on a stereo headset use this protocol to control the music player.

Audio/Video Distribution Transport Protocol

The *Audio/Video Distribution Transport Protocol* (AVDTP) is used by the advanced audio distribution profile to stream music to stereo headsets over an L2CAP channel intended for video distribution profile in the Bluetooth transmission.

Telephony Control Protocol

The *Telephony Control Protocol – Binary* (TCS BIN) is the bit-oriented protocol that defines the call control signaling for the establishment of voice and data calls between Bluetooth devices. Additionally, "TCS BIN defines mobility management procedures for handling groups of Bluetooth TCS devices."

TCS-BIN is only used by the cordless telephony profile, which failed to attract implementers. As such it is only of historical interest.

Adopted Protocols

Adopted protocols are defined by other standards-making organizations and incorporated into Bluetooth's protocol stack, allowing Bluetooth to code protocols only when necessary. The adopted protocols include:

Point-to-Point Protocol (PPP):

> Internet standard protocol for transporting IP datagrams over a point-to-point link.

TCP/IP/UDP:

> Foundation Protocols for TCP/IP protocol suite.

Object Exchange Protocol (OBEX):

> Session-layer protocol for the exchange of objects, providing a model for object and operation representation.

Wireless Application Environment/Wireless Application Protocol (WAE/WAP):

> WAE specifies an application framework for wireless devices and WAP is an open standard to provide mobile users access to telephony and information services.

Baseband Error Correction

Depending on packet type, individual packets may be protected by error correction, either 1/3 rate forward error correction (FEC) or 2/3 rate. In addition, packets with CRC will be retransmitted until acknowledged by automatic repeat request (ARQ).

Setting up Connections

Any Bluetooth device in *discoverable mode* transmits the following information on demand:

- Device name.
- Device class.
- List of services.
- Technical information (for example: device features, manufacturer, Bluetooth specification used, clock offset).

Any device may perform an inquiry to find other devices to connect to, and any device can be configured to respond to such inquiries. However, if the device trying to connect knows the address of the device, it always responds to direct connection requests and transmits the information shown in the list above if requested. Use of a device's services may require pairing or acceptance by its owner, but the connection itself can be initiated by any device and held until it goes out of range. Some devices can be connected to only one device at a time, and connecting to them prevents them from connecting to other devices and appearing in inquiries until they disconnect from the other device.

Every device has a unique 48-bit address. However, these addresses are generally not shown in inquiries. Instead, friendly Bluetooth names are used, which can be set by the user. This name appears when another user scans for devices and in lists of paired devices.

Most cellular phones have the Bluetooth name set to the manufacturer and model of the phone by default. Most cellular phones and laptops show only the Bluetooth names and special programs are required to get additional information about remote devices. This can be confusing as, for example, there could be several cellular phones in range named T610.

Pairing and Bonding

Motivation

Many services offered over Bluetooth can expose private data or let a connecting party control the Bluetooth device. Security reasons make it necessary to recognize specific devices, and thus enable control over which devices can connect to a given Bluetooth device. At the same time, it is useful for Bluetooth devices to be able to establish a connection without user intervention (for example, as soon as in range).

To resolve this conflict, Bluetooth uses a process called *bonding*, and a bond is generated through a process called *pairing*. The pairing process is triggered either by a specific request from a user to generate a bond (for example, the user explicitly requests to "Add a Bluetooth device"), or it is triggered automatically when connecting to a service where (for the first time) the identity of a device is required for security purposes. These two cases are referred to as dedicated bonding and general bonding respectively.

Pairing often involves some level of user interaction. This user interaction confirms the identity of the devices. When pairing successfully completes, a bond forms between the two devices, enabling those two devices to connect to each other in the future without repeating the pairing process to confirm device identities. When desired, the user can remove the bonding relationship.

Implementation

During pairing, the two devices establish a relationship by creating a shared secret known as a *link key*. If both devices store the same link key, they are said to be *paired* or *bonded*. A device that wants to communicate only with a bonded device can cryptographically authenticate the identity of the other device, ensuring it is the same device it previously paired with. Once a link key is generated, an authenticated Asynchronous Connection-Less (ACL) link between the devices may be encrypted to protect exchanged data against eavesdropping. Users can delete link keys from either device, which removes the bond between the devices—so it is possible for one device to have a stored link key for a device it is no longer paired with.

Bluetooth services generally require either encryption or authentication and as such require pairing before they let a remote device connect. Some services, such as the Object Push Profile, elect not to explicitly require authentication or encryption so that pairing does not interfere with the user experience associated with the service use-cases.

Pairing Mechanisms

Pairing mechanisms changed significantly with the introduction of Secure Simple Pairing in Bluetooth v2.1. The following summarizes the pairing mechanisms:

- *Legacy pairing*: This is the only method available in Bluetooth v2.0 and before. Each device must enter a PIN code; pairing is only successful if both devices enter the same PIN code. Any 16-byte UTF-8 string may be used as a PIN code; however, not all devices may be capable of entering all possible PIN codes.
    - *Limited input devices*: The obvious example of this class of device is a Bluetooth Hands-free headset, which generally have few inputs. These devices usually have a *fixed PIN*, for example "0000" or "1234", that are hard-coded into the device.
    - *Numeric input devices*: Mobile phones are classic examples of these devices. They allow a user to enter a numeric value up to 16 digits in length.
    - *Alpha-numeric input devices*: PCs and smartphones are examples of these devices. They allow a user to enter full UTF-8 text as a PIN code. If pairing with a less capable device the user must be aware of the input limitations on the other device; there is no mechanism available for a capable device to determine how it should limit the available input a user may use.
- *Secure Simple Pairing* (SSP): This is required by Bluetooth v2.1, although a Bluetooth v2.1 device may only use legacy pairing to interoperate with a v2.0 or earlier device. Secure Simple Pairing uses a form of public key cryptography, and some types can help protect against man in the middle, or MITM attacks. SSP has the following authentication mechanisms:
    - *Just works*: As the name implies, this method just works, with no user interaction. However, a device may prompt the user to confirm the pairing process. This method is typically used by headsets with very limited IO capabilities, and is more secure than the fixed PIN mechanism this limited set of devices uses for legacy pairing. This method provides no man-in-the-middle (MITM) protection.
    - *Numeric comparison*: If both devices have a display, and at least one can accept a binary yes/no user input, they may use Numeric Comparison. This method displays a 6-digit numeric code on each device. The user should compare the numbers to ensure they are identical. If the comparison succeeds, the user(s) should confirm pairing on the device(s) that can accept an input. This method provides MITM protection, assuming the user confirms on both devices and actually performs the comparison properly.

- *Passkey Entry*: This method may be used between a device with a display and a device with numeric keypad entry (such as a keyboard), or two devices with numeric keypad entry. In the first case, the display presents a 6-digit numeric code to the user, who then enters the code on the keypad. In the second case, the user of each device enters the same 6-digit number. Both of these cases provide MITM protection.
- *Out of band* (OOB): This method uses an external means of communication, such as near-field communication (NFC) to exchange some information used in the pairing process. Pairing is completed using the Bluetooth radio, but requires information from the OOB mechanism. This provides only the level of MITM protection that is present in the OOB mechanism.

SSP is considered simple for the following reasons:

- In most cases, it does not require a user to generate a passkey.
- For use cases not requiring MITM protection, user interaction can be eliminated.
- For *numeric comparison*, MITM protection can be achieved with a simple equality comparison by the user.
- Using OOB with NFC enables pairing when devices simply get close, rather than requiring a lengthy discovery process.

Security Concerns

Prior to Bluetooth v2.1, encryption is not required and can be turned off at any time. Moreover, the encryption key is only good for approximately 23.5 hours; using a single encryption key longer than this time allows simple XOR attacks to retrieve the encryption key.

- Turning off encryption is required for several normal operations, so it is problematic to detect if encryption is disabled for a valid reason or for a security attack.

Bluetooth v2.1 addresses this in the following ways:

- Encryption is required for all non-SDP (Service Discovery Protocol) connections.
- A new Encryption Pause and Resume feature is used for all normal operations that require that encryption be disabled. This enables easy identification of normal operation from security attacks.
- The encryption key must be refreshed before it expires.

Link keys may be stored on the device file system, not on the Bluetooth chip itself. Many Bluetooth chip manufacturers let link keys be stored on the device—however, if the device is removable, this means that the link key moves with the device.

Security

Bluetooth implements confidentiality, authentication and key derivation with custom algorithms based on the SAFER+ block cipher. Bluetooth key generation is generally based on a Bluetooth PIN, which must be entered into both devices. This procedure might be modified if one of the devices has a fixed PIN (e.g., for headsets or similar devices with a restricted user interface). During pairing, an initialization key or master key is generated, using the E22 algorithm. The E0 stream cipher is used for encrypting packets, granting confidentiality, and is based on a shared cryptographic secret, namely a previously generated link key or master key. Those keys, used for subsequent encryption of data sent via the air interface, rely on the Bluetooth PIN, which has been entered into one or both devices.

An overview of Bluetooth vulnerabilities exploits was published in 2007 by Andreas Becker.

In September 2008, the National Institute of Standards and Technology (NIST) published a Guide to Bluetooth Security as a reference for organizations. It describes Bluetooth security capabilities and how to secure Bluetooth technologies effectively. While Bluetooth has its benefits, it is susceptible to denial-of-service attacks, eavesdropping, man-in-the-middle attacks, message modification, and resource misappropriation. Users and organizations must evaluate their acceptable level of risk and incorporate security into the lifecycle of Bluetooth devices. To help mitigate risks, included in the NIST document are security checklists with guidelines and recommendations for creating and maintaining secure Bluetooth piconets, headsets, and smart card readers.

Bluetooth v2.1 – finalized in 2007 with consumer devices first appearing in 2009 – makes significant changes to Bluetooth's security, including pairing.

Bluejacking

Bluejacking is the sending of either a picture or a message from one user to an unsuspecting user through Bluetooth wireless technology. Common applications include short messages, e.g., "You've just been bluejacked!" Bluejacking does not involve the removal or alteration of any data from the device. Bluejacking can also involve taking control of a mobile device wirelessly and phoning a premium rate line, owned by the bluejacker. Security advances have alleviated this issue.

Health Concerns

Bluetooth uses the microwave radio frequency spectrum in the 2.402 GHz to 2.480 GHz range, which is non-ionizing radiation, of similar bandwidth to the one used by wireless and mobile phones. No specific demonstration of harm has been demonstrated up to date, even if wireless transmission has been included by IARC in the possible carcinogen list. Maximum power output from a Bluetooth radio is 100 mW for class 1, 2.5 mW for class 2, and 1 mW for class 3 devices. Even the maximum power output of class 1 is a lower level than the lowest-powered mobile phones. UMTS and W-CDMA output 250 mW, GSM1800/1900 outputs 1000 mW, and GSM850/900 outputs 2000 mW.

Wi-Fi

Wi-Fi is the name of a popular wireless networking technology that uses radio waves to provide wireless high-speed Internet and network connections. A common misconception is that the term Wi-Fi is short for "*wireless fidelity,*" however this is not the case. Wi-Fi is simply a trademarked phrase that means *IEEE 802.11x.*

Wi-Fi networks have no physical wired connection between sender and receiver by using radio frequency (RF) technology -- a frequency within the electromagnetic spectrum associated with radio wave propagation. When an RF current is supplied to an antenna, an electromagnetic field is created that then is able to propagate through space.

The cornerstone of any wireless network is an access point (AP). The primary job of an access point is to broadcast a wireless signal that computers can detect and "tune" into. In order to connect to an access point and join a wireless network, computers and devices must be equipped with wireless network adapters.

The Wi-Fi Alliance, the organization that owns the Wi-Fi registered trademark term specifically defines Wi-Fi as any "*wireless local area network (WLAN) products that are based on the Institute of Electrical and Electronics Engineers' (IEEE) 802.11 standards.*"

The old Wi-Fi Alliance logo

Introduced	September 1998; 19 years ago
Compatible hardware	Personal computers, gaming consoles, televisions, printers, mobile phones

Initially, Wi-Fi was used in place of only the 2.4GHz 802.11b standard, however the Wi-Fi Alliance has expanded the generic use of the Wi-Fi term to include any type of network or WLAN product based on any of the 802.11 standards, including 802.11b, 802.11a, dual-band and so on, in an attempt to stop confusion about wireless LAN interoperability.

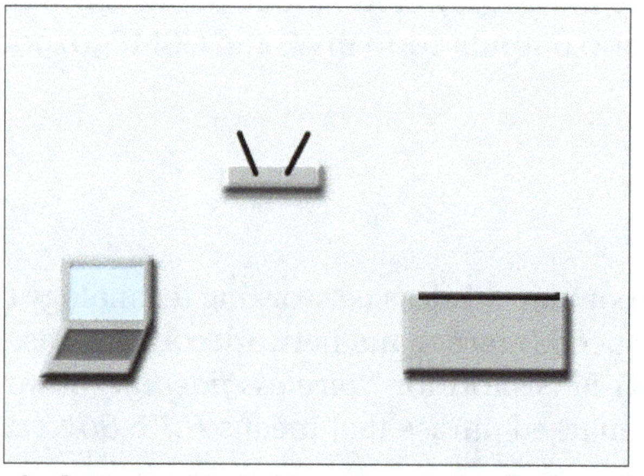

Depiction of a device sending information wirelessly to another device, both connected to the local network, in order to print a document.

Wi-Fi Certification

The IEEE does not test equipment for compliance with their standards. The non-profit Wi-Fi Alliance was formed in 1999 to fill this void — to establish and enforce standards for interoperability and backward compatibility, and to promote wireless local-area-network technology. As of 2010, the Wi-Fi Alliance consisted of more than 375 companies from around the world. The Wi-Fi Alliance enforces the use of the Wi-Fi brand to technologies based on the IEEE 802.11 standards from the IEEE. This includes wireless local area network (WLAN) connections, device to device connectivity (such as Wi-Fi Peer to Peer aka Wi-Fi Direct), Personal area network (PAN), local area network (LAN) and even some limited wide area network (WAN) connections. Manufacturers with membership in the Wi-Fi Alliance, whose products pass the certification process, gain the right to mark those products with the Wi-Fi logo.

Specifically, the certification process requires conformance to the IEEE 802.11 radio standards, the WPA2 and WPA3 security standards, and the EAP authentication standard. Certification may optionally include tests of IEEE 802.11 draft standards, interaction with cellular-phone technology in converged devices, and features relating to security set-up, multimedia, and power-saving.

Not every Wi-Fi device is submitted for certification. The lack of Wi-Fi certification does not necessarily imply that a device is incompatible with other Wi-Fi devices. The Wi-Fi Alliance may or may not sanction derivative terms, such as Super Wi-Fi, coined by the US Federal Communications Commission (FCC) to describe proposed networking in the UHF TV band in the US.

Terminology

A Japanese sticker indicating to the public that a location is within range of a Wi-Fi network. A dot with curved lines radiating from it is a common symbol for Wi-Fi, representing a point transmitting a signal.

To connect to a Wi-Fi LAN, a computer has to be equipped with a wireless network interface controller. The combination of computer and interface controllers is called a *station*. For all stations that share a single radio frequency communication channel, transmissions on this channel are received by all stations within range. The transmission is not guaranteed to be delivered and is therefore a best-effort delivery mechanism. A carrier wave is used to transmit the data. The data is organised in packets on an Ethernet link, referred to as "Ethernet frames".

A service set is the set of all the devices associated with a particular Wi-Fi network. The service set can be local, independent, extended or mesh.

Each service set has an associated identifier, the 32-byte Service Set Identifier (SSID), which identifies the particular network. The SSID is configured within the devices that are considered part of the network, and it is transmitted in the packets. Receivers ignore wireless packets from networks with a different SSID.

Uses

Internet Access

Wi-Fi technology may be used to provide Internet access to devices that are within the range of a wireless network that is connected to the Internet. The coverage of one or more interconnected access points (*hotspots*) can extend from an area as small as a few rooms to as large as many square kilometres. Coverage in the larger area may require a group of access points with overlapping coverage. For example, public outdoor Wi-Fi technology has been used successfully in wireless mesh networks in London, UK. An international example is Fon.

Wi-Fi provides service in private homes, businesses, as well as in public spaces at Wi-Fi hotspots set up either free-of-charge or commercially, often using a captive portal webpage for access. Organizations and businesses, such as airports, hotels, and restaurants, often provide free-use hotspots to attract customers. Enthusiasts or authorities who wish to provide services or even to promote business in selected areas sometimes provide free Wi-Fi access.

Routers that incorporate a digital subscriber line modem or a cable modem and a Wi-Fi access point, often set up in homes and other buildings, provide Internet access and internetworking to all devices connected to them, wirelessly or via cable.

Similarly, battery-powered routers may include a cellular Internet radio modem and Wi-Fi access point. When subscribed to a cellular data carrier, they allow nearby Wi-Fi stations to access the Internet over 2G, 3G, or 4G networks using the tethering technique. Many smartphones have a built-in capability of this sort, including those based on Android, BlackBerry, Bada, iOS (iPhone), Windows Phone and Symbian, though carriers often disable the feature, or charge a separate fee to enable it, especially for customers with unlimited data plans. "Internet packs" provide standalone facilities of this type as well, without use of a smartphone; examples include the MiFi- and WiBro-branded devices. Some laptops that have a cellular modem card can also act as mobile Internet Wi-Fi access points.

Wi-Fi also connects places that normally don't have network access, such as kitchens and garden sheds.

Google is intending to use the technology to allow rural areas to enjoy connectivity by utilizing a broad mix of projection and routing services. Google also intends to bring connectivity to Africa and some Asian lands by launching blimps that will allow for internet connection with Wi-Fi technology.

City-wide Wi-Fi

An outdoor Wi-Fi access point.

In the early 2000s, many cities around the world announced plans to construct city-wide Wi-Fi networks. There are many successful examples; in 2004, Mysore (Mysuru) became India's first Wi-Fi-enabled city. A company called WiFiyNet has set up hotspots in Mysore, covering the complete city and a few nearby villages.

In 2005, St. Cloud, Florida and Sunnyvale, California, became the first cities in the United States to offer citywide free Wi-Fi (from MetroFi). Minneapolis has generated $1.2 million in profit annually for its provider.

In May 2010, London mayor Boris Johnson pledged to have London-wide Wi-Fi by 2012. Several boroughs including Westminster and Islington already had extensive outdoor Wi-Fi coverage at that point.

Officials in South Korea's capital Seoul are moving to provide free Internet access at more than 10,000 locations around the city, including outdoor public spaces, major streets and densely populated residential areas. Seoul will grant leases to KT, LG Telecom, and SK Telecom. The companies will invest $44 million in the project, which was to be completed in 2015.

Campus-wide Wi-Fi

Many traditional university campuses in the developed world provide at least partial Wi-Fi coverage. Carnegie Mellon University built the first campus-wide wireless Internet network, called Wireless Andrew, at its Pittsburgh campus in 1993 before Wi-Fi branding originated. By February 1997 the CMU Wi-Fi zone was fully operational. Many universities collaborate in providing Wi-Fi access to students and staff through the Eduroam international authentication infrastructure.

Wi-Fi Ad Hoc versus Wi-Fi direct

Wi-Fi also allows communications directly from one computer to another without an access point intermediary. This is called *ad hoc* Wi-Fi transmission. This wireless ad hoc network mode has proven popular with multiplayer handheld game consoles, such as the Nintendo DS, PlayStation Portable, digital cameras, and other consumer electronics devices. Some devices can also share their Internet connection using ad hoc, becoming hotspots or "virtual routers".

A keychain-size Wi-Fi detector.

Similarly, the Wi-Fi Alliance promotes the specification Wi-Fi Direct for file transfers and media sharing through a new discovery- and security-methodology. Wi-Fi Direct launched in October 2010.

Another mode of direct communication over Wi-Fi is Tunneled Direct Link Setup (TDLS), which enables two devices on the same Wi-Fi network to communicate directly, instead of via the access point.

Wi-Fi Radio Spectrum

802.11b and 802.11g use the 2.4 GHz ISM band, operating in the United States under Part 15 Rules and Regulations. Because of this choice of frequency band, 802.11b and g equipment may occasionally suffer interference from microwave ovens, cordless telephones, USB3.0 hubs, and Bluetooth devices.

Spectrum assignments and operational limitations are not consistent worldwide: Australia and Europe allow for an additional two channels (12, 13) beyond the 11 permitted in the United States for the 2.4 GHz band, while Japan has three more (12–14). In the US and other countries, 802.11a and 802.11g devices may be operated without a license, as allowed in Part 15 of the FCC Rules and Regulations.

A Wi-Fi signal occupies five channels in the 2.4 GHz band. Any two channel numbers that differ by five or more, such as 2 and 7, do not overlap. The oft-repeated adage that channels 1, 6, and 11 are the *only* non-overlapping channels is, therefore, not accurate. Channels 1, 6, and 11 are the only *group of three* non-overlapping channels in North America and the United Kingdom. In Europe and Japan using Channels 1, 5, 9, and 13 for 802.11g and 802.11n is recommended.

802.11a uses the 5 GHz U-NII band, which, for much of the world, offers at least 23 non-overlapping channels rather than the 2.4 GHz ISM frequency band, where adjacent channels overlap.

Interference

Wi-Fi connections can be disrupted or the Internet speed lowered by having other devices in the same area. Wi-Fi protocols are designed to share channels reasonably fairly, and will often work with little to no disruption. However, many 2.4 GHz 802.11b and 802.11g access-points default to the same channel on initial startup, contributing to congestion on certain channels. Wi-Fi pollution, or an excessive number of access points in the area, can prevent access and interfere with other devices' use of other access points as well as with decreased signal-to-noise ratio (SNR) between access points. In addition interference can be caused by overlapping channels in the 802.11g/b spectrum. These issues can become a problem in high-density areas, such as large apartment complexes or office buildings with many Wi-Fi access points.

Additionally, other devices use the 2.4 GHz band: microwave ovens, ISM band devices, security cameras, ZigBee devices, Bluetooth devices, video senders, cordless phones, baby monitors, and, in some countries, amateur radio, all of which can cause significant additional interference. It is also an issue when municipalities or other large entities (such as universities) seek to provide large area coverage.

These bands are allowed to be used with low power transmitters, without requiring a license and with few restrictions. However, while unintended interference is common, users that have been found to knowingly cause deliberate interference to other users, particularly for attempting to locally monopolise these bands for commercial purposes, have been handed large fines.

Communications Protocol

USB wireless adapter.

The IEEE 802.11 standard is a set of media access control (MAC) and physical layer (PHY) specifications for implementing wireless local area network (WLAN) computer communication in the 2.4, 3.6, 5, and 60 GHz frequency bands. They are created and maintained by the IEEE LAN/MAN Standards Committee (IEEE 802). The base version of the standard was released in 1997, and has had subsequent amendments. The standard and amendments provide the basis for wireless network products using the Wi-Fi brand. While each amendment is officially revoked when it is incorporated in the latest version of the standard, the corporate world tends to market to the revisions because they concisely denote capabilities of their products. As a result, in the market place, each revision tends to become its own standard.

Range and Speed

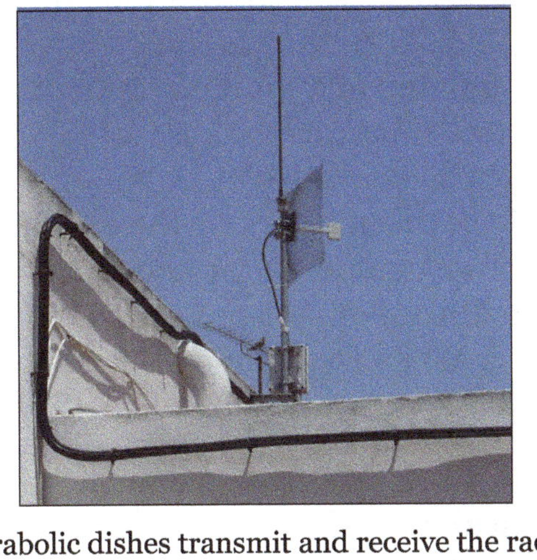

Parabolic dishes transmit and receive the radio waves only in particular directions and can give much greater range than omnidirection antennas.

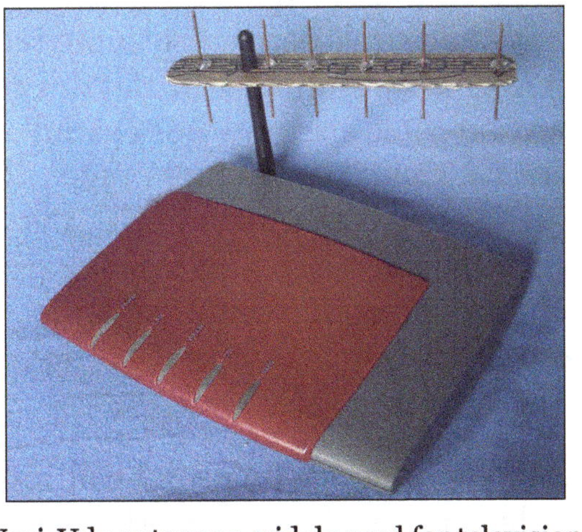

Yagi-Uda antennas, widely used for television reception, are relatively compact at WiFi wavelengths.

Wi-Fi operational range depends on factors such as the frequency band, radio power output, receiver sensitivity, antenna gain and antenna type as well as the modulation technique. In addition, propagation characteristics of the signals can have a big impact.

At longer distances, and with greater signal absorption, speed is usually reduced.

Transmitter Power

Compared to cell phones and similar technology, Wi-Fi transmitters are low power devices. In general, the maximum amount of power that a Wi-Fi device can transmit is limited by local regulations, such as FCC Part 15 in the US. Equivalent isotropically radiated power (EIRP) in the European Union is limited to 20 dBm (100 mW).

To reach requirements for wireless LAN applications, Wi-Fi has higher power consumption compared to some other standards designed to support wireless personal area network (PAN) applications. For example Bluetooth provides a much shorter propagation range between 1 and 100 m and so in general have a lower power consumption. Other low-power technologies such as ZigBee have fairly long range, but much lower data rate. The high power consumption of Wi-Fi makes battery life in some mobile devices a concern.

Antenna

An access point compliant with either 802.11b or 802.11g, using the stock omnidirectional antenna might have a range of 100 m (0.062 mi). The same radio with an external semi parabolic antenna (15 dB gain) with a similarly equipped receiver at the far end might have a range over 20 miles.

Higher gain rating (dBi) indicates further deviation (generally toward the horizontal) from a theoretical, perfect isotropic radiator, and therefore the antenna can project a usable signal further in particular directions, as compared to a similar output power on a more isotropic antenna. For example, an 8 dBi antenna used with a 100 mW driver will have a similar horizontal range to a 6 dBi antenna being driven at 500 mW. Note that this assumes that radiation in the vertical is lost; this may not be the case in some situations, especially in large buildings or within a waveguide. In the above example, a directional waveguide could cause the low power 6 dBi antenna to project much further in a single direction than the 8 dBi antenna which is not in a waveguide, even if they are both being driven at 100 mW.

On wireless routers with detachable antennas, it is possible to improve range by fitting upgraded antennas which have higher gain in particular directions. Outdoor ranges can be improved to many kilometers through the use of high gain directional antennas at the router and remote device(s).

MIMO

This Netgear Wi-Fi router contains dual bands for transmitting the 802.11 standard across the 2.4 and 5 GHz spectrums and supports MIMO.

Some standards, such as IEEE 802.11n and IEEE 802.11ac for Wi-Fi allow a device to have multiple antennas. Multiple antennas enable the equipment to focus on the far end device, reducing interference in other directions, and giving a stronger useful signal. This greatly increases range and network speed without exceeding the legal power limits.

IEEE802.11n can more than double the range. Range also varies with frequency band. Wi-Fi in the 2.4 GHz frequency block has slightly better range than Wi-Fi in the 5 GHz frequency block used by 802.11a (and optionally by 802.11n).

Under optimal conditions, IEEE802.11ac can achieve communication rates of 1Gb/s.

Alternatives

For the best performance, a number of people only recommend using wireless networking as a supplement to wired networking.

Researchers have developed a number of "no new wires" technologies to provide alternatives to Wi-Fi for applications in which Wi-Fi's indoor range is not adequate and where installing new wires (such as CAT-6) is not possible or cost-effective. For example, the ITU-T G.hn standard for high speed local area networks uses existing home wiring (coaxial cables, phone lines and power lines). Although G.hn does not provide some of the advantages of Wi-Fi (such as mobility or outdoor use), it is designed for applications (such as IPTV distribution) where indoor range is more important than mobility.

Many mobile devices also employ the cell phone networks to provide access to the Internet when out of range of Wi-Fi networks. This is done, rather than rely entirely on cell networks as Wi-Fi networks are often cheaper for bulk data usage.

Radio Propagation

With Wi-Fi signals line-of-sight usually works best, but Wi-Fi signals can be affected by absorption, reflection, and diffraction through and around structures.

Due to the complex nature of radio propagation at typical Wi-Fi frequencies, particularly the effects of signal reflection off trees and buildings, algorithms can only approximately predict Wi-Fi signal strength for any given area in relation to a transmitter. This effect does not apply equally to long-range Wi-Fi, since longer links typically operate from towers that transmit above the surrounding foliage.

The practical range of Wi-Fi essentially confines mobile use to such applications as inventory-taking machines in warehouses or in retail spaces, barcode-reading devices at check-out stands, or receiving/shipping stations. Mobile use of Wi-Fi over wider ranges is limited, for instance, to uses such as in an automobile moving from one hotspot to another. Other wireless technologies are more suitable for communicating with moving vehicles.

Distance Records

Distance records (using non-standard devices) include 382 km (237 mi) in June 2007, held by Ermanno Pietrosemoli and EsLaRed of Venezuela, transferring about 3 MB of data between the mountain-tops of El Águila and Platillon. The Swedish Space Agency transferred data 420 km (260 mi), using 6 watt amplifiers to reach an overhead stratospheric balloon.

Throughput

As the 802.11 specifications evolved to support higher throughput, the bandwidth requirements also increased to support them. 802.11n uses double the radio spectrum/bandwidth (40 MHz) compared to 802.11a or 802.11g (20 MHz).76 This means there can be only one 802.11n network on the 2.4 GHz band at a given location, without interference to/from other WLAN traffic. 802.11n can also be set to limit itself to 20 MHz bandwidth to prevent interference in dense community.

Many newer consumer devices support the latest 802.11ac standard, which uses the 5 GHz band exclusively and is capable of multi-station WLAN throughput of at least 1 gigabit per second, and a single station throughput of at least 500 Mbit/s. In the first quarter of 2016, The Wi-Fi Alliance certifies devices compliant with the 802.11ac standard as "Wi-Fi CERTIFIED ac". This new standard uses several advanced signal processing techniques such as multi-user MIMO and 4X4 Spatial Multiplexing streams, and large channel bandwidth (160 MHz) to achieve the Gigabit throughput. According to a study by IHS Technology, 70% of all access point sales revenue In the first quarter of 2016 came from 802.11ac devices.

Multiple Access Points

Increasing the number of Wi-Fi access points for a network provides redundancy, better range, support for fast roaming and increased overall network-capacity by using more channels or by defining smaller cells. Except for the smallest implementations (such as home or small office networks), Wi-Fi implementations have moved toward "thin" access points, with more of the network intelligence housed in a centralized network appliance, relegating individual access points to the role of "dumb" transceivers. Outdoor applications may use mesh topologies.

When multiple access points are deployed they are often configured with the same SSID and any security settings to form an "extended service set". Wi-Fi client devices will typically connect to the access point that can provide the strongest signal within that service set.

Hardware

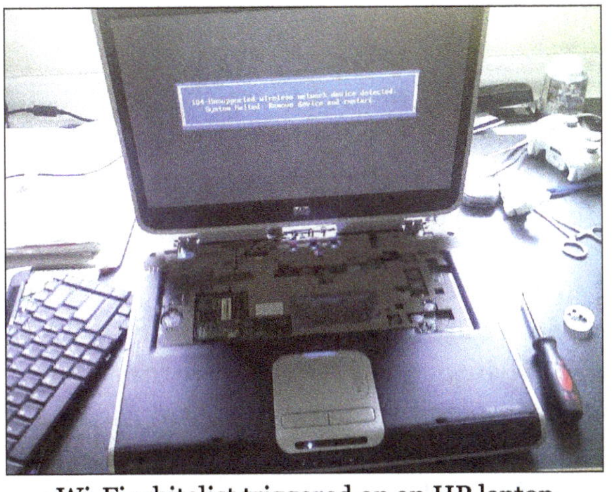

Wi-Fi whitelist triggered on an HP laptop.

Wi-Fi allows wireless deployment of local area networks (LANs). Also, spaces where cables cannot be run, such as outdoor areas and historical buildings, can host wireless LANs. However, building walls of certain materials, such as stone with high metal content, can block Wi-Fi signals.

Since early 2000's manufacturers are building wireless network adapters into most laptops. The price of chipsets for Wi-Fi continues to drop, making it an economical networking option included in even more devices.

Different competitive brands of access points and client network-interfaces can inter-operate at a basic level of service. Products designated as "Wi-Fi Certified" by the Wi-Fi Alliance are backward compatible. Unlike mobile phones, any standard Wi-Fi device will work anywhere in the world.

Standard Devices

A wireless access point (WAP) connects a group of wireless devices to an adjacent wired LAN. An access point resembles a network hub, relaying data between connected wireless devices in addition to a (usually) single connected wired device, most often an Ethernet hub or switch, allowing wireless devices to communicate with other wired devices.

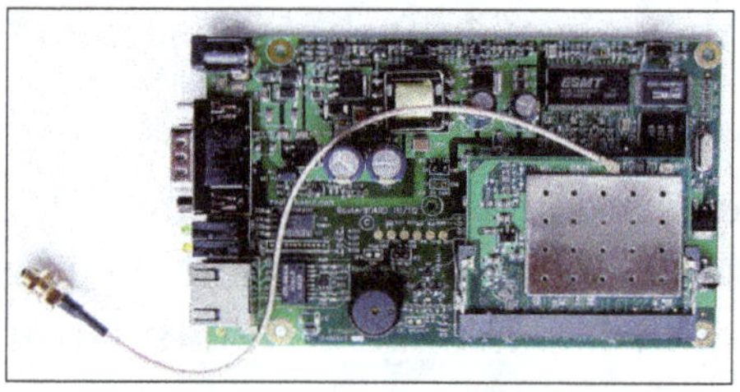

An embedded RouterBoard 112 with U.FL-RSMA pigtail and R52 mini PCI Wi-Fi card widely used by wireless Internet service providers (WISPs) in the Czech Republic.

OSBRiDGE 3GN – 802.11n Access Point and UMTS/GSM Gateway in one device.

An Atheros draft-N Wi-Fi adapter with built in Bluetooth on a laptop.

Wireless adapters allow devices to connect to a wireless network. These adapters connect to devices using various external or internal interconnects such as PCI, miniPCI, USB, ExpressCard, Cardbus and PC Card. As of 2010, most newer laptop computers come equipped with built in internal adapters.

Wireless routers integrate a Wireless Access Point, Ethernet switch, and internal router firmware application that provides IP routing, NAT, and DNS forwarding through an integrated WAN-interface. A wireless router allows wired and wireless Ethernet LAN devices to connect to a (usually) single WAN device such as a cable modem or a DSL modem. A wireless router allows all three devices, mainly the access point and router, to be configured through one central utility. This utility is usually an integrated web server that is accessible to wired and wireless LAN clients and often optionally to WAN clients. This utility may also be an application that is run on a computer, as is the case with as Apple's AirPort, which is managed with the AirPort Utility on macOS and iOS.

Wireless network bridges connect a wired network to a wireless network. A bridge differs from an access point: an access point connects wireless devices to a wired network at the data-link layer. Two wireless bridges may be used to connect two wired networks over a wireless link, useful in situations where a wired connection may be unavailable, such as between two separate homes or for devices which do not have wireless networking capability (but have wired networking capability), such as consumer entertainment devices; alternatively, a wireless bridge can be used to enable a device which supports a wired connection to operate at a wireless networking standard which is faster than supported by the wireless network connectivity feature (external dongle or inbuilt) supported by the device (e.g. enabling Wireless-N speeds (up to the maximum supported speed on the wired Ethernet port on both the bridge and connected devices including the wireless access point) for a device which only supports Wireless-G). A dual-band wireless bridge can also be used to enable 5 GHz wireless network operation on a device which only supports 2.4 GHz wireless networking functionality and has a wired Ethernet port.

Wireless range-extenders or wireless repeaters can extend the range of an existing wireless network. Strategically placed range-extenders can elongate a signal area or allow for the signal area to reach around barriers such as those pertaining in L-shaped corridors. Wireless devices connected through repeaters will suffer from an increased latency for each hop, as well as from a reduction in the maximum data throughput that is available. In addition, the effect of additional users using a network employing wireless range-extenders is to consume the available bandwidth faster than would be the case where but a single user migrates around a network employing extenders. For this reason, wireless range-extenders work best in networks supporting very low traffic throughput requirements, such as for cases where but a single user with a Wi-Fi equipped tablet migrates around the combined extended and non-extended portions of the total connected network. Additionally, a wireless device connected to any of the repeaters in the chain will have a data throughput that is also limited by the "weakest link" existing in the chain between where the connection originates and where the connection ends. Networks employing wireless extenders are also more prone to degradation from interference from neighboring access points that border portions of the extended network and that happen to occupy the same channel as the extended network.

The security standard, Wi-Fi Protected Setup, allows embedded devices with limited graphical user interface to connect to the Internet with ease. Wi-Fi Protected Setup has 2 configurations: The Push Button configuration and the PIN configuration. These embedded devices are also called The Internet of Things and are low-power, battery-operated embedded systems. A number of Wi-Fi manufacturers design chips and modules for embedded Wi-Fi, such as GainSpan.

Embedded Systems

Embedded serial-to-Wi-Fi module.

Increasingly in the last few years (particularly as of 2007), embedded Wi-Fi modules have become available that incorporate a real-time operating system and provide a simple means of wirelessly enabling any device which has and communicates via a serial port. This allows the design of simple monitoring devices. An example is a portable

ECG device monitoring a patient at home. This Wi-Fi-enabled device can communicate via the Internet.

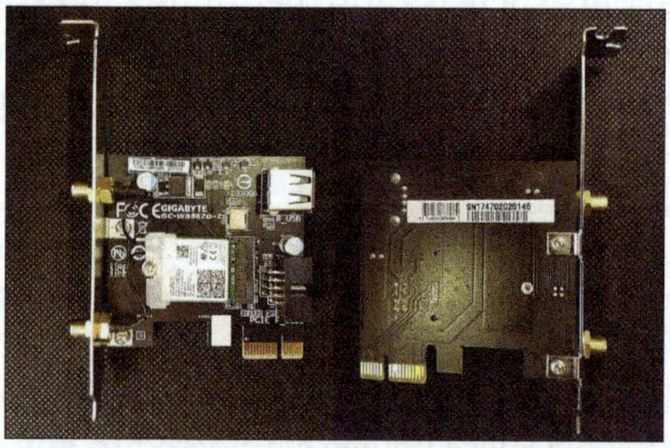

Wireless network interface controller Gigabyte GC-WB867D-I.

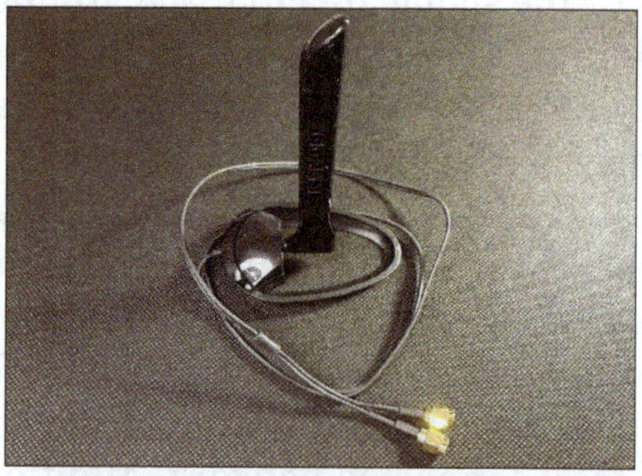

Antenna of wireless network interface controller Gigabyte GC-WB867D-I.

These Wi-Fi modules are designed by OEMs so that implementers need only minimal Wi-Fi knowledge to provide Wi-Fi connectivity for their products.

In June 2014, Texas Instruments introduced the first ARM Cortex-M4 microcontroller with an onboard dedicated Wi-Fi MCU, the SimpleLink CC3200. It makes embedded systems with Wi-Fi connectivity possible to build as single-chip devices, which reduces their cost and minimum size, making it more practical to build wireless-networked controllers into inexpensive ordinary objects.

Network Security

The main issue with wireless network security is its simplified access to the network compared to traditional wired networks such as Ethernet. With wired networking, one must either gain access to a building (physically connecting into the internal network), or break through an external firewall. To enable Wi-Fi, one merely needs to be within the range of the Wi-Fi network. Most business networks protect sensitive data and systems by attempting to disallow external access. Enabling wireless connectivity reduces security if the network uses inadequate or no encryption.

An attacker who has gained access to a Wi-Fi network router can initiate a DNS spoofing attack against any other user of the network by forging a response before the queried DNS server has a chance to reply.

Securing Methods

A common measure to deter unauthorized users involves hiding the access point's name by disabling the SSID broadcast. While effective against the casual user, it is ineffective as a security method because the SSID is broadcast in the clear in response to a client SSID query. Another method is to only allow computers with known MAC addresses to join the network, but determined eavesdroppers may be able to join the network by spoofing an authorized address.

Wired Equivalent Privacy (WEP) encryption was designed to protect against casual snooping but it is no longer considered secure. Tools such as AirSnort or Aircrack-ng can quickly recover WEP encryption keys. Because of WEP's weakness the Wi-Fi Alliance approved Wi-Fi Protected Access (WPA) which uses TKIP. WPA was specifically designed to work with older equipment usually through a firmware upgrade. Though more secure than WEP, WPA has known vulnerabilities.

The more secure WPA2 using Advanced Encryption Standard was introduced in 2004 and is supported by most new Wi-Fi devices. WPA2 is fully compatible with WPA. In 2017 a flaw in the WPA2 protocol was discovered, allowing a key replay attack, known as KRACK.

A flaw in a feature added to Wi-Fi in 2007, called Wi-Fi Protected Setup (WPS), allows WPA and WPA2 security to be bypassed and effectively broken in many situations. The only remedy as of late 2011 is to turn off Wi-Fi Protected Setup, which is not always possible.

Virtual Private Networks are often used to secure Wi-Fi.

Data Security Risks

The older wireless encryption-standard, Wired Equivalent Privacy (WEP), has been shown to be easily breakable even when correctly configured. Wi-Fi Protected Access (WPA and WPA2) encryption, which became available in devices in 2003, aimed to solve this problem. Wi-Fi access points typically default to an encryption-free (*open*) mode. Novice users benefit from a zero-configuration device that works out-of-the-box, but this default does not enable any wireless security, providing open wireless access to a LAN. To turn security on requires the user to configure the device, usually via a software graphical user interface (GUI). On unencrypted Wi-Fi networks connecting devices can monitor and record data (including personal information). Such networks can only be secured by using other means of protection, such as a VPN or secure Hypertext Transfer Protocol over Transport Layer Security (HTTPS).

Wi-Fi Protected Access encryption (WPA2) is considered secure, provided a strong passphrase is used. In 2018, WPA3 was announced as a replacement for WPA2, increasing security; it rolled out on June 26.

Piggybacking

Piggybacking refers to access to a wireless Internet connection by bringing one's own computer within the range of another's wireless connection, and using that service without the subscriber's explicit permission or knowledge.

During the early popular adoption of 802.11, providing open access points for anyone within range to use was encouraged to cultivate wireless community networks, particularly since people on average use only a fraction of their downstream bandwidth at any given time.

Recreational logging and mapping of other people's access points has become known as wardriving. Indeed, many access points are intentionally installed without security turned on so that they can be used as a free service. Providing access to one's Internet connection in this fashion may breach the Terms of Service or contract with the ISP. These activities do not result in sanctions in most jurisdictions; however, legislation and case law differ considerably across the world. A proposal to leave graffiti describing available services was called warchalking. A Florida court case determined that owner laziness was not to be a valid excuse.

Piggybacking often occurs unintentionally – a technically unfamiliar user might not change the default "unsecured" settings to their access point and operating systems can be configured to connect automatically to any available wireless network. A user who happens to start up a laptop in the vicinity of an access point may find the computer has joined the network without any visible indication. Moreover, a user intending to join one network may instead end up on another one if the latter has a stronger signal. In combination with automatic discovery of other network resources this could possibly lead wireless users to send sensitive data to the wrong middle-man when seeking a destination. For example, a user could inadvertently use an unsecure network to log into a website, thereby making the login credentials available to anyone listening, if the website uses an unsecure protocol such as plain HTTP without TLS (HTTPS).

An unauthorized user can obtain security information (factory preset passphrase and/or Wi-Fi Protected Setup PIN) from a label on a wireless access point can use this information (or connect by the Wi-Fi Protected Setup pushbutton method) to commit unauthorized and/or unlawful activities.

Infrared Communication

Infrared (IR) is a wireless mobile technology used for device communication over short ranges. IR communication has major limitations because it requires line-of-sight, has a short transmission range and is unable to penetrate walls. IR transceivers are quite cheap and serve as short-range communication solutions.

Because of IR's limitations, communication interception is difficult. In fact, Infrared Data Association (IrDA) device communication is usually exchanged on a one-to-one basis. Thus, data transmitted between IrDA devices is normally unencrypted.

IR-enabled devices are known as IrDA devices because they conform to standards set by the Infrared Data Association (IrDA). IR light-emitting diodes (LED) are used to transmit IR signals, which pass through a lens and focus into a beam of IR data. The beam source is rapidly switched on and off for data encoding.

The IR beam data is received by an IrDA device equipped with a silicon photodiode. This receiver converts the IR beam into an electric current for processing. Because IR transitions more slowly from ambient light than from a rapidly pulsating IrDA signal, the silicon photodiode can filter out the IrDA signal from ambient IR.

IrDA transmitters and receivers are classified as directed and non-directed. A transmitter or receiver that uses a focused and narrow beam is directed, whereas a transmitter or receiver that uses an omnidirectional radiation pattern is non-directed.

Infrared band of the electromagnet corresponds to 430THz to 300GHz and a wavelength of **980nm**. The propagation of light waves in this band can be used for a communication system (for transmission and reception) of data. This communication can be between two portable devices or between a portable device and a fixed device.

There are two types of Infrared communication:

- Point to Point: It requires a line of sight between the transmitter and a receiver. In other words the transmitter and the receiver should be pointed to each other and there shouldn't be any obstacles between them. Example is the remote control communication.
- Diffuse Point: It doesn't require any line of sight and the link between the transmitter and the receiver is maintained by reflecting or bouncing of the transmitted signal by surfaces like ceilings, roof, etc. Example is the wireless LAN communication system

Advantages of IR communication:

- Security: Infrared communication has high directionality and can identify the source as different sources emit radiation of different frequencies and thus the risk of information being diffused is eliminated.
- Safety: Infrared radiation is not harmful to human beings. Hence infrared communication can be used at any place.
- High Speed data Communication: The data rate of Infrared communication is about 1Gbps and can be used for sending information like video signal.

IR Communication Basics:

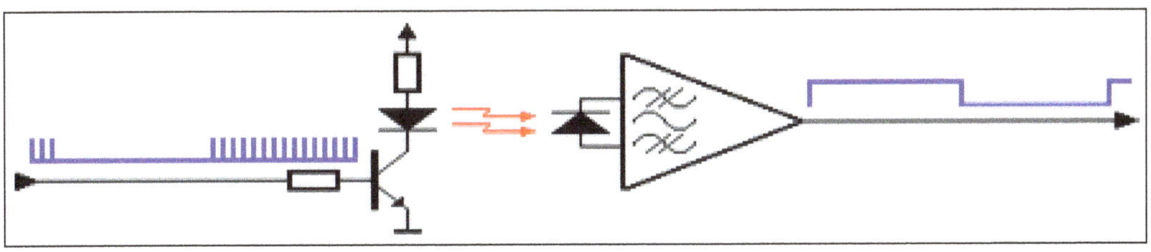

IR communication principle.

IR Transmission

The transmitter of an IR LED inside its circuit, which emits infrared light for every electric pulse given to it. This pulse is generated as a button on the remote is pressed, thus completing the circuit, providing bias to the LED.

The LED on being biased emits light of the wavelength of 940nm as a series of pulses, corresponding to the button pressed. However since along with the IR LED many other sources of infrared light such as us human beings, light bulbs, sun, etc, the transmitted information can be interfered. A solution to this problem is by modulation. The transmitted signal is modulated using a carrier frequency of 38 KHz (or any other frequency between 36 to 46 KHz). The IR LED is made to oscillate at this frequency for the time duration of the pulse. The information or the light signals are pulse width modulated and are contained in the 38 KHz frequency.

IR Reception

The receiver consists of a photodetector which develops an output electrical signal as light is incident on it. The output of the detector is filtered using a narrow band filter that discards all the frequencies below or above the carrier frequency (38 KHz in this case). The filtered output is then given to the suitable device like a Microcontroller or a Microprocessor which controls devices like a PC or a Robot. The output from the filters can also be connected to the Oscilloscope to read the pulses.

Parts of IR Communication System:

IR Transmittor-IR Sensor

The sensors could be utilized as a part of measuring the radiation temperature without any contact. For different radiation temperature ranges various filters are available. An infrared (IR) sensor is an electronic device that radiates or locates infrared radiation to sense some part of its surroundings. They are undetectable to human eyes.

An infrared sensor could be considered a Polaroid that briefly recalls how an area's infrared radiation shows up. It is very regular for an infrared sensor to be coordinated into movement indicators like those utilized as a feature of private or business security systems. An IR sensor is basically has two terminals positive and negative. These sensors are undetectable to human eyes. They can measure the heat of an object and also identify movement. The region wavelength roughly from 0.75μm to 1000 μm is the IR region. The wavelength region of 0.75μm to 3 μm is called close infrared, the region from 3 μm to 6 μm is called mid infrared and the region higher than 6 μm is called far infrared. IR sensors emits at a frequency of 38 KHz.

Features of IR Sensor:

- Input voltage: 5VDC

- Sensing Range: 5cm
- Output signal: analog voltage
- Emitting element: Infrared LED

Example Interfacing Circuit of IR Diode and Photodiode

IR sensors mostly used in radiation thermometer, gas analyzers, industrial applications, IR imaging devices, tracking, and human body detection, communication and health hazards

An IR diode is connected through a resistance to the dc supply. A photo diode is connected in reverse biased condition through a potential divider of a 10k variable resistance and 1k in series to the base of the transistor. While the IR rays fall on the reverse biased photo diode it conducts that causes a voltage at the base of the transistor.

The transistor then works like a switch while the collector goes to ground. Once the IR rays are obstructed the driving voltage is not available to the transistor thus its collector goes high. This low to high logic can be used for the microcontroller input for any action as per the program.

IR Receiver/TSOP Sensor – Features & Specifications

TSOP is the standard IR remote control receiver series, supporting all major transmission codes. This is capable of receiving infrared radiation modulated at 38 kHz. IR sensors we have seen up to now working just for little short distance up to 6 cm. TSOP is sensitive to a specific frequency so its range is better contrast with ordinary photo diode. We can alter it up to 15 cm.

TSOP acts like as a receiver. It has three pins GND, Vs and OUT. GND is connected to common ground, Vs is connected to +5volts and OUT is connected to output pin. TSOP sensor has an inbuilt control circuit for amplifying the coded pulses from the IR transmitter. These are commonly used in TV remote receivers. As I said above TSOP sensors sense only a particular frequency.

Features:

- The preamplifier and photo detector both are in single package.
- Internal filter for PCM frequency.
- Improved shielding against electrical field disturbance.
- TTL and CMOS compatibility.
- Output active low.

- Low power consumption.
- High immunity against ambient light.
- Continuous data transmission possible.

Specifications:

- Supply Voltage is –0.3-6.0 V
- Supply Current is 5 mA
- Output Voltage is –0.3-6.0 V
- Output Current is 5 mA
- Storage Temperature Range is –25-+85 °C
- Operating Temperature Range is –25-+85°C

The testing of TSOP is very simple. These are commonly used in TV remote receivers. TSOP consists of a PIN diode and pre-amplifier internally. Connect TSOP sensor as shown in circuit. A LED is connected through a resistance from the supply to output.

And then when we press the button of T.V. Remote control in front of the TSOP sensor, if LED starts blinking then our TSOP sensor and its connection is correct. The point when the output of TSOP is low i.e. at the time it appropriates IR signal from a source, with a centre frequency of 38 kHz, its output goes low.

TSOP sensor is used in our daily use TV, VCD, music system's remote control. Where IR rays are transmitted by pushing a button on remote which are received by TSOP receiver inside the equipment.

References

- Muller, Scott. Upgrading and Reparing PCs (22nd ed.). Pearson Education. p. 898. ISBN 978-0-7897-5610-7
- "Reaching the Mobile Respondent: Determinants of High-Level Mobile Phone Use Among a High-Coverage Group" (PDF). Social Science Computer Review. doi:10.1177/0894439309353099
- Lean, Geoffrey; Shawcross, Harriet (April 15, 2007). "Are mobile phones wiping out our bees?". The Independent. UK. Archived from the original on July 6, 2008. Retrieved May 12, 2010
- "Rise in executions for mobile use". ITV News. June 15, 2007. Archived from the original on August 17, 2007. Retrieved June 23, 2007
- Marziah Karch (1 September 2010). Android for Work: Productivity for Professionals. Apress. ISBN 978-1-4302-3000-7. Archived from the original on 17 February 2013. Retrieved 11 November 2012
- Laberge-Nadeau, Claire (September 2003). "Wireless telephones and the risk of road crashes". Accident Analysis & Prevention. 35 (5): 649–660. doi:10.1016/S0001-4575(02)00043-X

- Kurawar, Arwa; Koul, Ayushi; Patil, Viki Tukaram (August 2014). "Survey of Bluetooth and Applications". International Journal of Advanced Research in Computer Engineering & Technology. 3: 2832–2837. ISSN 2278-1323
- Wolter Lemstra; Vic Hayes; John Groenewegen (2010). The innovation journey of Wi-Fi: the road to global success. Cambridge University Press. p. 121. ISBN 978-0-521-19971-1
- The Committee Office, House of Commons. "Supplementary memorandum submitted by Gregory Smith". Publications.parliament.uk. Retrieved 2011-07-11

5

Radio Communication and Technology

Radio is a technology to send electromagnetic signals to deliver information from one place to another. Radio propagation is the behavior of radiowaves when they are transmitted from one point to another. The chapter closely examines the key concepts of radios to provide an extensive understanding of the subject.

Radio

Radio is the technology of using radio waves to carry information, such as sound, by systematically modulating some property of electromagnetic energy waves transmitted through space, such as their amplitude, frequency, phase, or pulse width. When radio waves strike an electrical conductor, the oscillating fields induce an alternating current in the conductor. The information in the waves can be extracted and transformed back into its original form.

The Alexandra Palace, here: mast of the broadcasting station.

Radio systems need a transmitter to modulate (change) some property of the energy produced to impress a signal on it, for example using amplitude modulation or angle modulation (which can be frequency modulation or phase modulation). Radio systems also need an antenna to convert electric currents into radio waves, and vice versa. An antenna can be used for both transmitting and receiving. The electrical resonance of tuned circuits in radios allow individual stations to be selected. The electromagnetic wave is intercepted by a tuned receiving antenna. A radio receiver receives its input from an antenna and converts it into a form usable for the consumer, such as sound, pictures, digital data, measurement values, navigational positions, etc. Radio frequencies occupy the range from a 3 kHz to 300 GHz, although commercially important uses of radio use only a small part of this spectrum.

Classic radio receiver dial.

A radio communication system sends signals by radio. The radio equipment involved in communication systems includes a transmitter and a receiver, each having an antenna and appropriate terminal equipment such as a microphone at the transmitter and a loudspeaker at the receiver in the case of a voice-communication system.

Etymology

Following Heinrich Hertz's establishment of the existence of electromagnetic radiation in the late 1880s, a variety of terms were initially used for the phenomenon, with early descriptions of the radiation itself including "Hertzian waves", "electric waves", and "ether waves", while phrases describing its use in communications included "spark telegraphy", "space telegraphy", "aerography" and, eventually and most commonly, "wireless telegraphy". However, "wireless" included a broad variety of related electronic technologies, including electrostatic induction, electromagnetic induction and aquatic and earth conduction, so there was a need for a more precise term referring exclusively to electromagnetic radiation.

The first use of *radio-* in conjunction with electromagnetic radiation appears to have been by French physicist Édouard Branly, who in 1890 developed a version of a coherer receiver he called a *radio-conducteur*. The radio- prefix was later used to form additional descriptive compound and hyphenated words, especially in Europe, for example, in early 1898 the British publication *The Practical Engineer* included a reference to "the radiotelegraph" and "radiotelegraphy", while the

French text of both the 1903 and 1906 Berlin Radiotelegraphic Conventions includes the phrases *radiotélégraphique* and *radiotélégrammes.*

The use of "radio" as a standalone word dates back to at least December 30, 1904, when instructions issued by the British Post Office for transmitting telegrams specified that "The word 'Radio'... is sent in the Service Instructions". This practice was universally adopted, and the word "radio" introduced internationally, by the 1906 Berlin Radiotelegraphic Convention, which included a Service Regulation specifying that "Radiotelegrams shall show in the preamble that the service is 'Radio'".

The switch to "radio" in place of "wireless" took place slowly and unevenly in the English-speaking world. Lee de Forest helped popularize the new word in the United States—in early 1907 he founded the DeForest Radio Telephone Company, and his letter in the June 22, 1907 *Electrical World* about the need for legal restrictions warned that "Radio chaos will certainly be the result until such stringent regulation is enforced". The United States Navy would also play a role. Although its translation of the 1906 Berlin Convention used the terms "wireless telegraph" and "wireless telegram", by 1912 it began to promote the use of "radio" instead. The term started to become preferred by the general public in the 1920s with the introduction of broadcasting. ("Broadcasting" is based upon an agricultural term meaning roughly "scattering seeds widely".) British Commonwealth countries continued to commonly use the term "wireless" until the mid-20th century, though the magazine of the British Broadcasting Corporation in the UK has been called Radio Times since its founding in the early 1920s.

In recent years the more general term "wireless" has gained renewed popularity, even for devices using electromagnetic radiation, through the rapid growth of short-range computer networking, e.g., Wireless Local Area Network (WLAN), Wi-Fi, and Bluetooth, as well as mobile telephony, e.g., GSM and UMTS cell phones. Today, the term "radio" specifies the transceiver device or chip, whereas "wireless" refers to the lack of physical connections; thus equipment employs embedded *radio* transceivers, but operates as *wireless* devices over *wireless* sensor networks.

Processes

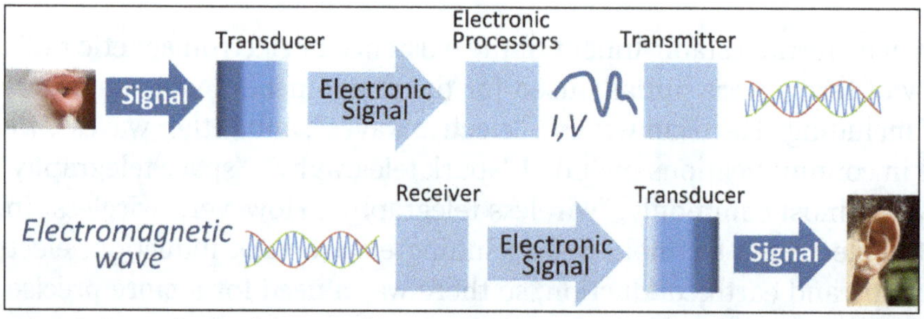

Radio communication. Information such as sound is converted by a transducer such as a microphone to an electrical signal, which modulates a radio wave sent from a transmitter. A receiver intercepts the radio wave and extracts the information-bearing electronic signal, which is converted back using another transducer such as a speaker.

Radio systems used for communication have the following elements. With more than 100 years of development, each process is implemented by a wide range of methods, specialised for different communications purposes.

Transmitter and Modulation

Each system contains a transmitter, This consists of a source of electrical energy, producing alternating current of a desired frequency of oscillation. The transmitter contains a system to modulate (change) some property of the energy produced to impress a signal on it. This modulation might be as simple as turning the energy on and off, or altering more subtle properties such as amplitude, frequency, phase, or combinations of these properties. The transmitter sends the modulated electrical energy to a tuned resonant antenna; this structure converts the rapidly changing alternating current into an electromagnetic wave that can move through free space (sometimes with a particular polarization).

Amplitude modulation of a carrier wave works by varying the strength of the transmitted signal in proportion to the information being sent. For example, changes in the signal strength can be used to reflect the sounds to be reproduced by a speaker, or to specify the light intensity of television pixels. It was the method used for the first audio radio transmissions, and remains in use today. "AM" is often used to refer to the medium wave broadcast band, but it is used in various radiotelephone services such as the Citizen Band, amateur radio and especially in aviation, due to its ability to be received under very weak signal conditions and its immunity to capture effect, allowing more than one signal to be heard simultaneously.

Frequency modulation varies the frequency of the carrier. The instantaneous frequency of the carrier is directly proportional to the instantaneous value of the input signal. FM has the "capture effect" whereby a receiver only receives the strongest signal, even when others are present. Digital data can be sent by shifting the carrier's frequency among a set of discrete values, a technique known as frequency-shift keying. FM is commonly used at Very high frequency (VHF) radio frequencies for high-fidelity broadcasts of music and speech. Analog TV sound is also broadcast using FM.

Angle modulation alters the instantaneous phase of the carrier wave to transmit a signal. It may be either FM or phase modulation (PM).

Antenna

Rooftop television antennas. Yagi-Uda antennas like these six are widely used at VHF and UHF frequencies.

An *antenna* (or *aerial*) is an electrical device which converts electric currents into radio waves, and vice versa. It is usually used with a radio transmitter or radio receiver. In transmission, a radio transmitter supplies an electric current oscillating at radio frequency (i.e. high frequency AC) to the antenna's terminals, and the antenna radiates the energy from the current as electromagnetic waves (radio waves). In reception, an antenna intercepts some of the power of an electromagnetic wave in order to produce a tiny voltage at its terminals, that is applied to a receiver to be amplified. Some antennas can be used for both transmitting and receiving, even simultaneously, depending on the connected equipment.

Propagation

Once generated, electromagnetic waves travel through space either directly, or have their path altered by reflection, refraction or diffraction. The intensity of the waves diminishes due to geometric dispersion (the inverse-square law); some energy may also be absorbed by the intervening medium in some cases. Noise will generally alter the desired signal; this electromagnetic interference comes from natural sources, as well as from artificial sources such as other transmitters and accidental radiators. Noise is also produced at every step due to the inherent properties of the devices used. If the magnitude of the noise is large enough, the desired signal will no longer be discernible; the signal-to-noise ratio is the fundamental limit to the range of radio communications.

Resonance

Electrical resonance of tuned circuits in radios allow individual stations to be selected. A resonant circuit will respond strongly to a particular frequency, and much less so to differing frequencies. This allows the radio receiver to discriminate between multiple signals differing in frequency.

Receiver and Demodulation

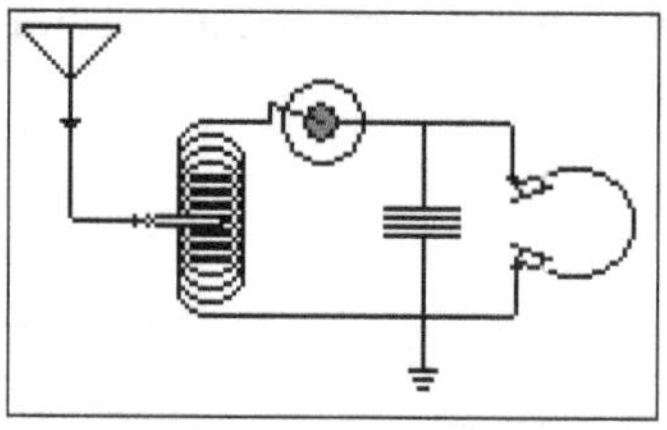

A crystal receiver, consisting of an antenna, adjustable electromagnetic coil, crystal rectifier, capacitor, headphones and ground connection.

The electromagnetic wave is intercepted by a tuned receiving antenna; this structure captures some of the energy of the wave and returns it to the form of oscillating electrical currents. At the receiver, these currents are demodulated, which is conversion to a usable signal form by a detector sub-system. The receiver is "tuned" to respond preferentially to the desired signals, and reject undesired signals.

Early radio systems relied entirely on the energy collected by an antenna to produce signals for the operator. Radio became more useful after the invention of electronic devices such as the vacuum tube and later the transistor, which made it possible to amplify weak signals. Today radio systems are used for applications from walkie-talkie children's toys to the control of space vehicles, as well as for broadcasting, and many other applications.

A *radio receiver* receives its input from an antenna, uses electronic filters to separate a wanted radio signal from all other signals picked up by this antenna, amplifies it to a level suitable for further processing, and finally converts through demodulation and decoding the signal into a form usable for the consumer, such as sound, pictures, digital data, measurement values, navigational positions, etc.

Radio Band

Light comparison		
Name	**Frequency (Hz) (Wavelength)**	**Photon energy (eV)**
Gamma ray	> 30 EHz (0.01 nm)	124 keV - 300+ GeV
X-Ray	30 EHz - 30 PHz (0.01 nm - 10 nm)	124 eV to 120 keV
Ultraviolet	30 PHz - 750 THz (10 nm - 400 nm)	3.1 eV to 124 eV
Visible	750 THz - 428.5 THz (400 nm - 700 nm)	1.7 eV - 3.1 eV
Infrared	428.5 THz - 300 GHz (700 nm - 1 mm)	1.24 meV - 1.7 eV
Microwave	300 GHz - 300 MHz (1 mm - 1 m)	1.24 μeV - 1.24 meV
Radio	300 MHz - 3 kHz (1 m - 100 km)	12.4 feV - 1.24 meV

Radio frequencies occupy the range from a 3 kHz to 300 GHz, although commercially important uses of radio use only a small part of this spectrum. Other types of electromagnetic radiation, with frequencies above the RF range, are infrared, visible light, ultraviolet, X-rays and gamma rays. Since the energy of an individual photon of radio frequency is too low to remove an electron from an atom, radio waves are classified as non-ionizing radiation.

Communication Systems

A *radio communication system* sends signals by radio. Types of radio communication systems deployed depend on technology, standards, regulations, radio spectrum allocation, user requirements, service positioning, and investment.

The radio equipment involved in communication systems includes a transmitter and a receiver, each having an antenna and appropriate terminal equipment such as a microphone at the transmitter and a loudspeaker at the receiver in the case of a voice-communication system.

The power consumed in a transmitting station varies depending on the distance of communication and the transmission conditions. The power received at the receiving station is usually only a tiny fraction of the transmitter's output, since communication depends on receiving the information, not the energy, that was transmitted.

Classical radio communications systems use frequency-division multiplexing (FDM) as a strategy to split up and share the available radio-frequency bandwidth for use by different parties communications concurrently. Modern radio communication systems include those that divide up a radio-frequency band by time-division multiplexing (TDM) and code-division multiplexing (CDM) as alternatives to the classical FDM strategy. These systems offer different tradeoffs in supporting multiple users, beyond the FDM strategy that was ideal for broadcast radio but less so for applications such as mobile telephony.

A radio communication system may send information only one way. For example, in broadcasting a single transmitter sends signals to many receivers. Two stations may take turns sending and receiving, using a single radio frequency; this is called "simplex." By using two radio frequencies, two stations may continuously and concurrently send and receive signals - this is called "duplex" operation.

History

In 1864 James Clerk Maxwell showed mathematically that electromagnetic waves could propagate through free space. The effects of electromagnetic waves (then-unexplained "action at a distance" sparking behavior) were actually observed before and after Maxwell's work by many inventors and experimenters including Luigi Galvani (1791), Peter Samuel Munk (1835), Joseph Henry (1842), Samuel Alfred Varley (1852), Edwin Houston, Elihu Thomson, Thomas Edison (1875) and David Edward Hughes (1878). Edison gave the effect the name "etheric force" and Hughes detected a spark impulse up to 500 yards (460 m) with a portable receiver, but none could identify what caused the phenomenon and it was usually written off as electromagnetic induction. In 1886 Heinrich Rudolf Hertz noticed the same sparking phenomenon and, in published experiments (1887-1888), was able to demonstrate the existence of electromagnetic waves in an experiment confirming Maxwell's theory of electromagnetism. The discovery of these "Hertzian waves" (radio waves) prompted many experiments by physicists. An August 1894 lecture by the British physicist Oliver Lodge, where he transmitted and received "Hertzian waves" at distances up to 50 meters, was followed up a year later with experiments by Indian physicist Jagadish Bose in radio microwave optics and construction of a radio based lightning detector by Russian physicist Alexander Stepanovich Popov. Starting in late 1894, Guglielmo Marconi began pursuing the idea of building a wireless telegraphy system based on Hertzian waves (radio). Marconi gained a patent on the system in 1896 and developed it into a commercial communication system over the next few years.

Early 20th century radio systems transmitted messages by continuous wave code only. Early attempts at developing a system of amplitude modulation for voice and music were demonstrated in 1900 and 1906, but had little success. World War I accelerated the development of radio for military communications, and in this era the first vacuum tubes were applied to radio transmitters and receivers. Electronic amplification was a key development in changing radio from an experimental practice by experts into a home appliance. After the war, commercial radio broadcasting began in the 1920s and became an important mass medium for entertainment and news.

World War II again accelerated development of radio for the wartime purposes of aircraft and land communication, radio navigation and radar. After the war, the experiments in television that had been interrupted were resumed, and it also became an important home entertainment medium.

Uses of Radio

Early uses were maritime, for sending telegraphic messages using Morse code between ships and land. The earliest users included the Japanese Navy scouting the Russian fleet during the Battle of Tsushima in 1905. One of the most memorable uses of marine telegraphy was during the sinking of the RMS *Titanic* in 1912, including communications between operators on the sinking ship and nearby vessels, and communications to shore stations listing the survivors.

Radio was used to pass on orders and communications between armies and navies on both sides in World War I; Germany used radio communications for diplomatic messages once it discovered that its submarine cables had been tapped by the British. The United States passed on President Woodrow Wilson's Fourteen Points to Germany via radio during the war. Broadcasting began from San Jose, California in 1909, and became feasible in the 1920s, with the widespread introduction of radio receivers, particularly in Europe and the United States. Besides broadcasting, point-to-point broadcasting, including telephone messages and relays of radio programs, became widespread in the 1920s and 1930s. Another use of radio in the pre-war years was the development of detection and locating of aircraft and ships by the use of radar (*RA*dio *D*etection *A*nd *R*anging).

Today, radio takes many forms, including wireless networks and mobile communications of all types, as well as radio broadcasting. Before the advent of television, commercial radio broadcasts included not only news and music, but dramas, comedies, variety shows, and many other forms of entertainment (the era from the late 1920s to the mid-1950s is commonly called radio's "Golden Age"). Radio was unique among methods of dramatic presentation in that it used only sound.

Audio

One-way

Bakelite radio at the Bakelite Museum, Orchard Mill, Williton, Somerset, UK.

AM radio uses amplitude modulation, in which the amplitude of the transmitted signal is made proportional to the sound amplitude captured (transduced) by the microphone, while the transmitted frequency remains unchanged. Transmissions are affected by static and interference because lightning and other sources of radio emissions on the same frequency add their amplitudes to the original transmitted amplitude.

A Fisher 500 AM/FM hi-fi receiver from 1959.

In the early part of the 20th century, American AM radio stations broadcast with powers as high as 500 kW, and some could be heard worldwide; these stations' transmitters were commandeered for military use by the US Government during World War II. Currently, the maximum broadcast power for a civilian AM radio station in the United States and Canada is 50 kW, and the majority of stations that emit signals this powerful were grandfathered in. In 1986 KTNN received the last granted 50,000-watt class A license. These 50 kW stations are generally called "clear channel" stations, because within North America each of these stations has exclusive use of its broadcast frequency throughout part or all of the broadcast day.

Bush House, old home of the BBC World Service.

FM broadcast radio sends music and voice with less noise than AM radio. It is often mistakenly thought that FM is higher fidelity than AM, but that is not true. AM is capable of the same audio bandwidth that FM employs. AM receivers typically use narrower filters in the receiver to recover the signal with less noise. AM stereo receivers can reproduce the same audio bandwidth that FM does due to the wider filter used in an AM stereo receiver, but today, AM radios limit the audio bandpass to 3–5 kHz. In frequency modulation, amplitude variation at the microphone causes the transmitter frequency to fluctuate. Because the audio signal modulates the frequency and not the amplitude, an FM signal is not subject to static and interference in the same way as AM signals. Due to its need for a wider bandwidth, FM is transmitted in the Very High Frequency (VHF, 30 MHz to 300 MHz) radio spectrum.

VHF radio waves act more like light, traveling in straight lines; hence the reception range is generally limited to about 50–200 miles (80–322 km). During unusual upper atmospheric conditions, FM signals are occasionally reflected back towards the Earth by the ionosphere, resulting in long distance FM reception. FM receivers are subject to the capture effect, which causes the radio to only receive the strongest signal when multiple signals appear on the same frequency. FM receivers are relatively immune to lightning and spark interference.

High power is useful in penetrating buildings, diffracting around hills, and refracting in the dense atmosphere near the horizon for some distance beyond the horizon. Consequently, 100,000-watt FM stations can regularly be heard up to 100 miles (160 km) away, and farther, 150 miles (240 km), if there are no competing signals.

A few old, "grandfathered" stations do not conform to these power rules. WBCT-FM (93.7) in Grand Rapids, Michigan, US, runs 320,000 watts ERP, and can increase to 500,000 watts ERP by the terms of its original license. Such a huge power level does not usually help to increase range as much as one might expect, because VHF frequencies travel in nearly straight lines over the horizon and off into space. Nevertheless, when there were fewer FM stations competing, this station could be heard near Bloomington, Illinois, US, almost 300 miles (480 km) away.

FM subcarrier services are secondary signals transmitted in a "piggyback" fashion along with the main program. Special receivers are required to utilize these services. Analog channels may contain alternative programming, such as reading services for the blind, background music or stereo sound signals. In some extremely crowded metropolitan areas, the sub-channel program might be an alternate foreign-language radio program for various ethnic groups. Sub-carriers can also transmit digital data, such as station identification, the current song's name, web addresses, or stock quotes. In some countries, FM radios automatically re-tune themselves to the same channel in a different district by using sub-bands.

Two-way

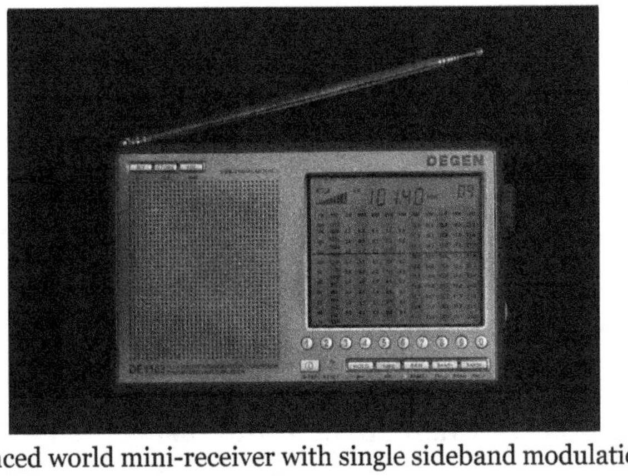

Degen DE1103, an advanced world mini-receiver with single sideband modulation and dual conversion.

Aviation voice radios use Aircraft band VHF AM. AM is used so that multiple stations on the same channel can be received. (Use of FM would result in stronger stations blocking out reception of weaker stations due to FM's capture effect). Aircraft fly high enough that their transmitters can be received hundreds of miles away, even though they are using VHF.

Marine voice radios can use single sideband voice (SSB) in the shortwave High Frequency (HF—3 MHz to 30 MHz) radio spectrum for very long ranges or Marine VHF radio / *narrowband FM* in the VHF spectrum for much shorter ranges. Narrowband FM sacrifices fidelity to make more channels available within the radio spectrum, by using a smaller range of radio frequencies, usually with five kHz of deviation, versus the 75 kHz used by commercial FM broadcasts, and 25 kHz used for TV sound.

Government, police, fire and commercial voice services also use narrowband FM on special frequencies. Early police radios used AM receivers to receive one-way dispatches.

Civil and military HF (high frequency) voice services use shortwave radio to contact ships at sea, aircraft and isolated settlements. Most use single sideband voice (SSB), which uses less bandwidth than AM. On an AM radio SSB sounds like ducks quacking, or the adults in a Charlie Brown cartoon. Viewed as a graph of frequency versus power, an AM signal shows power where the frequencies of the voice add and subtract with the main radio frequency. SSB cuts the bandwidth in half by suppressing the carrier and one of the sidebands. This also makes the transmitter about three times more powerful, because it doesn't need to transmit the unused carrier and sideband.

TETRA, Terrestrial Trunked Radio is a digital cell phone system for military, police and ambulances. Commercial services such as XM, WorldSpace and Sirius offer encrypted digital satellite radio.

Telephony

Mobile phones transmit to a local cell site (transmitter/receiver) that ultimately connects to the public switched telephone network (PSTN) through an optic fiber or microwave radio and other network elements. When the mobile phone nears the edge of the cell site's radio coverage area, the central computer switches the phone to a new cell. Cell phones originally used FM, but now most use various digital modulation schemes. Recent developments in Sweden (such as DROPme) allow for the instant downloading of digital material from a radio broadcast (such as a song) to a mobile phone.

Satellite phones use satellites rather than cell towers to communicate.

Video

Analog television sends the picture as AM and the sound as AM or FM, with the sound carrier a fixed frequency (4.5 MHz in the NTSC system) away from the video carrier. Analog television also uses a vestigial sideband on the video carrier to reduce the bandwidth required.

Digital television uses 8VSB modulation in North America (under the ATSC digital television standard), and COFDM modulation elsewhere in the world (using the DVB-T standard). A Reed–Solomon error correction code adds redundant correction codes and allows reliable reception during moderate data loss. Although many current and future codecs can be sent in the MPEG transport stream container format, as of 2006 most systems use a standard-definition format almost identical to DVD: MPEG-2 video in Anamorphic widescreen and MPEG layer 2 (*MP2*) audio. High-definition television is possible simply by using a higher-resolution picture, but H.264/AVC is being considered as a replacement video codec in some regions for its improved compression. With the compression and improved modulation involved, a single "channel" can contain a high-definition program and several standard-definition programs.

Navigation

All satellite navigation systems use satellites with precision clocks. The satellite transmits its position, and the time of the transmission. The receiver listens to four satellites, and can figure its position as being on a line that is tangent to a spherical shell around each satellite, determined by the time-of-flight of the radio signals from the satellite. A computer in the receiver does the math.

Radio direction-finding is the oldest form of radio navigation. Before 1960 navigators used movable loop antennas to locate commercial AM stations near cities. In some cases they used marine radiolocation beacons, which share a range of frequencies just above AM radio with amateur radio operators. LORAN systems also used time-of-flight radio signals, but from radio stations on the ground.

Very High Frequency Omnidirectional Range (VOR), systems (used by aircraft), have an antenna array that transmits two signals simultaneously. A directional signal rotates like a lighthouse at a fixed rate. When the directional signal is facing north, an omnidirectional signal pulses. By measuring the difference in phase of these two signals, an aircraft can determine its bearing or radial from the station, thus establishing a line of position. An aircraft can get readings from two VORs and locate its position at the intersection of the two radials, known as a "fix."

When the VOR station is collocated with DME (Distance Measuring Equipment), the aircraft can determine its bearing and range from the station, thus providing a fix from only one ground station. Such stations are called VOR/DMEs. The military operates a similar system of navaids, called TACANs, which are often built into VOR stations. Such stations are called VORTACs. Because TACANs include distance measuring equipment, VOR/DME and VORTAC stations are identical in navigation potential to civil aircraft.

Radar

Radar (Radio Detection And Ranging) detects objects at a distance by bouncing radio waves off them. The delay caused by the echo measures the distance. The direction of the beam determines the direction of the reflection. The polarization and frequency of the return can sense the type of surface. Navigational radars scan a wide area two to four times per minute. They use very short waves that reflect from earth and stone. They are common on commercial ships and long-distance commercial aircraft.

General purpose radars generally use navigational radar frequencies, but modulate and polarize the pulse so the receiver can determine the type of surface of the reflector. The best general-purpose radars distinguish the rain of heavy storms, as well as land and vehicles. Some can superimpose sonar data and map data from GPS position.

Search radars scan a wide area with pulses of short radio waves. They usually scan the area two to four times a minute. Sometimes search radars use the Doppler effect to separate moving vehicles from clutter. Targeting radars use the same principle as search radar but scan a much smaller area far more often, usually several times a second or more. Weather radars resemble search radars, but use radio waves with circular polarization and a wavelength to reflect from water droplets. Some weather radar use the Doppler effect to measure wind speeds.

Data (Digital Radio)

2008 Pure One Classic digital radio.

Most new radio systems are digital, including Digital TV, satellite radio, and Digital Audio Broadcasting. The oldest form of digital broadcast was spark gap telegraphy, used by pioneers such as Marconi. By pressing the key, the operator could send messages in Morse code by energizing a rotating commutating spark gap. The rotating commutator produced a tone in the receiver, where a simple spark gap would produce a hiss, indistinguishable from static. Spark-gap transmitters are now illegal, because their transmissions span several hundred megahertz. This is very wasteful of both radio frequencies and power.

The next advance was continuous wave telegraphy, or CW (Continuous Wave), in which a pure radio frequency, produced by a vacuum tube electronic oscillator was switched on and off by a key. A receiver with a local oscillator would “heterodyne” with the pure radio frequency, creating a whistle-like audio tone. CW uses less than 100 Hz of bandwidth. CW is still used, these days primarily by amateur radio operators (hams). Strictly, on-off keying of a carrier should be known as “Interrupted Continuous Wave” or ICW or on-off keying (OOK).

Radioteletype equipment usually operates on short-wave (HF) and is much loved by the military because they create written information without a skilled operator. They send a bit as one of two tones using frequency-shift keying. Groups of five or seven bits become a character printed by a teleprinter. From about 1925 to 1975, radioteletype was how most commercial messages were sent to less developed countries. These are still used by the military and weather services.

Aircraft use a 1200 Baud radioteletype service over VHF to send their ID, altitude and position, and get gate and connecting-flight data. Microwave dishes on satellites, telephone exchanges and

TV stations usually use quadrature amplitude modulation (QAM). QAM sends data by changing both the phase and the amplitude of the radio signal. Engineers like QAM because it packs the most bits into a radio signal when given an exclusive (non-shared) fixed narrowband frequency range. Usually the bits are sent in "frames" that repeat. A special bit pattern is used to locate the beginning of a frame.

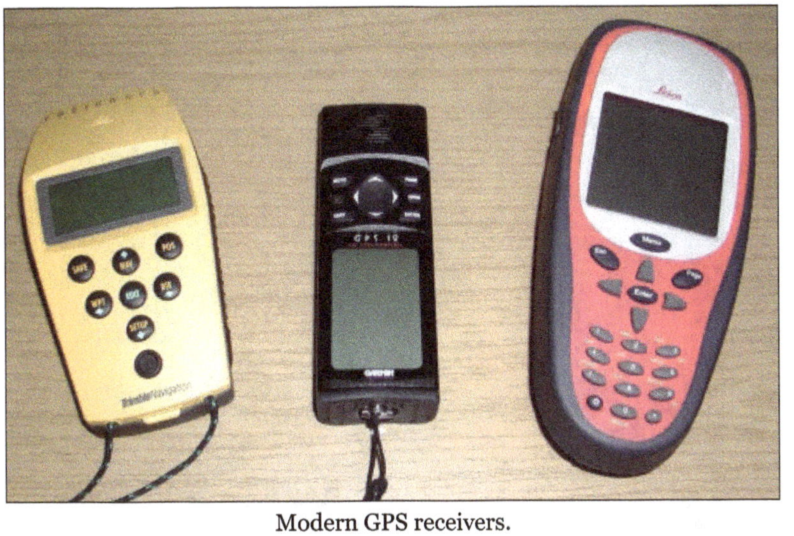

Modern GPS receivers.

Communication systems that limit themselves to a fixed narrowband frequency range are vulnerable to jamming. A variety of jamming-resistant spread spectrum techniques were initially developed for military use, most famously for Global Positioning System satellite transmissions. Commercial use of spread spectrum began in the 1980s. Bluetooth, most cell phones, and the 802.11b version of Wi-Fi each use various forms of spread spectrum.

Systems that need reliability, or that share their frequency with other services, may use "coded orthogonal frequency-division multiplexing" or COFDM. COFDM breaks a digital signal into as many as several hundred slower subchannels. The digital signal is often sent as QAM on the subchannels. Modern COFDM systems use a small computer to make and decode the signal with digital signal processing, which is more flexible and far less expensive than older systems that implemented separate electronic channels.

COFDM resists fading and ghosting because the narrow-channel QAM signals can be sent slowly. An adaptive system, or one that sends error-correction codes can also resist interference, because most interference can affect only a few of the QAM channels. COFDM is used for Wi-Fi, some cell phones, Digital Radio Mondiale, Eureka 147, and many other local area network, digital TV and radio standards.

Heating

Radio-frequency energy generated for heating of objects is generally not intended to radiate outside of the generating equipment, to prevent interference with other radio signals. Microwave ovens use intense radio waves to heat food. Diathermy equipment is used in surgery for sealing of blood vessels. Induction furnaces are used for melting metal for casting, and induction hobs for cooking.

Amateur Radio Service

Amateur radio, also known as "ham radio", is a hobby in which enthusiasts are licensed to communicate on a number of bands in the radio frequency spectrum non-commercially and for their own experiments. They may also provide emergency and service assistance in exceptional circumstances. This contribution has been very beneficial in saving lives in many instances.

Amateur radio station with multiple receivers and transceivers.

Radio amateurs use a variety of modes, including efficient ones like Morse code and experimental ones like Low-Frequency Experimental Radio. Several forms of radio were pioneered by radio amateurs and later became commercially important, including FM, single-sideband (SSB), AM, digital packet radio and satellite repeaters. Some amateur frequencies may be disrupted illegally by power-line internet service.

Unlicensed Radio Services

Unlicensed, government-authorized personal radio services such as Citizens' band radio in Australia, most of the Americas, and Europe, and Family Radio Service and Multi-Use Radio Service in North America exist to provide simple, usually short range communication for individuals and small groups, without the overhead of licensing. Similar services exist in other parts of the world. These radio services involve the use of handheld units.

Wi-Fi also operates in unlicensed radio bands and is very widely used to network computers.

Free radio stations, sometimes called pirate radio or "clandestine" stations, are unauthorized, unlicensed, illegal broadcasting stations. These are often low power transmitters operated on sporadic schedules by hobbyists, community activists, or political and cultural dissidents. Some pirate stations operating offshore in parts of Europe and the United Kingdom more closely resembled legal stations, maintaining regular schedules, using high power, and selling commercial advertising time.

Radio Control (RC)

Radio remote controls use radio waves to transmit control data to a remote object as in some early forms of guided missile, some early TV remotes and a range of model boats, cars and airplanes. Large industrial remote-controlled equipment such as cranes and switching locomotives now usually use digital radio techniques to ensure safety and reliability.

In Madison Square Garden, at the Electrical Exhibition of 1898, Nikola Tesla successfully demonstrated a radio-controlled boat. He was awarded U.S. patent No. 613,809 for a "Method of and Apparatus for Controlling Mechanism of Moving Vessels or Vehicles."

Radio Propagation

Radio propagation is the behavior of radio waves as they travel, or are propagated, from one point to another, or into various parts of the atmosphere. As a form of electromagnetic radiation, like light waves, radio waves are affected by the phenomena of reflection, refraction, diffraction, absorption, polarization, and scattering.

Radio propagation is affected by the daily changes of water vapor in the troposphere and ionization in the upper atmosphere, due to the Sun. Understanding the effects of varying conditions on radio propagation has many practical applications, from choosing frequencies for international shortwave broadcasters, to designing reliable mobile telephone systems, to radio navigation, to operation of radar systems.

Radio propagation is also affected by several other factors determined by its path from point to point. This path can be a direct line of sight path or an over-the-horizon path aided by refraction in the ionosphere, which is a region between approximately 60 and 600 km above the earth's surface. Factors influencing ionospheric radio signal propagation can include sporadic-E, spread-F, solar flares, geomagnetic storms, ionospheric layer tilts, and solar proton events.

Radio waves at different frequencies propagate in different ways. At extremely low frequencies (ELF) and very low frequencies the wavelength is much larger than the separation between the earth's surface and the D layer of the ionosphere, so electromagnetic waves may propagate in this region as a waveguide. Indeed, for frequencies below 20 kHz, the wave propagates as a single waveguide mode with a horizontal magnetic field and vertical electric field. The interaction of radio waves with the ionized regions of the atmosphere makes radio propagation more complex to predict and analyze than in free space. Ionospheric radio propagation has a strong connection to space weather. A sudden ionospheric disturbance or shortwave fadeout is observed when the x-rays associated with a solar flare ionize the ionospheric D-region. Enhanced ionization in that region increases the absorption of radio signals passing through it. During the strongest solar x-ray flares, complete absorption of virtually all ionospherically propagated radio signals in the sunlit hemisphere can occur. These solar flares can disrupt HF radio propagation and affect GPS accuracy.

Since radio propagation is not fully predictable, such services as emergency locator transmitters, in-flight communication with ocean-crossing aircraft, and some television broadcasting have been moved to communications satellites. A satellite link, though expensive, can offer highly predictable and stable line of sight coverage of a given area.

Free Space Propagation

In free space, all electromagnetic waves (radio, light, X-rays, etc.) obey the inverse-square law which states that the power density of an electromagnetic wave is proportional to the inverse of the square of the distance from a point source.

Doubling the distance from a transmitter means that the power density of the radiated wave at that new location is reduced to one-quarter of its previous value.

The power density per surface unit is proportional to the product of the electric and magnetic field strengths. Thus, doubling the propagation path distance from the transmitter reduces each of their received field strengths over a free-space path by one-half.

Modes

Radio frequencies and their primary mode of propagation				
Band		Frequency	Wavelength	Propagation via
ELF	Extremely Low Frequency	3–30 Hz	10,000-100,000 km	
SLF	Super Low Frequency	30–300 Hz	10,000-1,000 km	
ULF	Ultra Low Frequency	0.3–3 kHz (300–3,000 Hz)	1,000–100 km	
VLF	Very Low Frequency	3–30 kHz (3,000–30,000 Hz)	100–10 km	Guided between the earth and the ionosphere.
LF	Low Frequency	30–300 kHz (30,000–300,000 Hz)	10–1 km	Guided between the earth and the D layer of the ionosphere. Surface waves.
MF	Medium Frequency	300–3000 kHz (300,000–3,000,000 Hz)	1000–100 m	Surface waves. E, F layer ionospheric refraction at night, when D layer absorption weakens.
HF	High Frequency (Short Wave)	3–30 MHz (3,000,000–30,000,000 Hz)	100–10 m	E layer ionospheric refraction. F1, F2 layer ionospheric refraction.
VHF	Very High Frequency	30–300 MHz (30,000,000–300,000,000 Hz)	10–1 m	Infrequent E ionospheric (E_s) refraction. Uncommonly F2 layer ionospheric refraction during high sunspot activity up to 50 MHz and rarely to 80 MHz. Generally direct wave. Sometimes tropospheric ducting or meteor scatter

UHF	Ultra High Frequency	300–3000 MHz (300,000,000–3,000,000,000 Hz)	100–10 cm	Direct wave. Sometimes tropospheric ducting.
SHF	Super High Frequency	3–30 GHz (3,000,000,000–30,000,000,000 Hz)	10–1 cm	Direct wave. Sometimes rain scatter.
EHF	Extremely High Frequency	30–300 GHz (30,000,000,000–300,000,000,000 Hz)	10–1 mm	Direct wave limited by absorption.
THF	Tremendously High frequency	0.3–3 THz (300,000,000,000–3,000,000,000,000 Hz)	1–0.1 mm	

Surface Modes (Groundwave)

Lower frequencies (between 30 and 3,000 kHz) have the property of following the curvature of the earth via -[groundwave] propagation in the majority of occurrences.

In this mode the radio wave propagates by interacting with the semi-conductive surface of the earth. The wave "clings" to the surface and thus follows the curvature of the earth. Vertical polarization is used to alleviate short circuiting the electric field through the conductivity of the ground. Since the ground is not a perfect electrical conductor, ground waves are attenuated rapidly as they follow the earth's surface. Attenuation is proportional to the frequency making this mode mainly useful for LF and VLF frequencies.

Today LF and VLF are mostly used for time signals, and for military communications, especially one-way transmissions to ships and submarines, although radio amateurs have an allocation at 137 kHz in some parts of the world. Radio broadcasting using surface wave propagation uses the higher portion of the LF range in Europe, Africa and the Middle East.

Early commercial and professional radio services relied exclusively on long wave, low frequencies and ground-wave propagation. To prevent interference with these services, amateur and experimental transmitters were restricted to the higher (HF) frequencies, felt to be useless since their ground-wave range was limited. Upon discovery of the other propagation modes possible at medium wave and short wave frequencies, the advantages of HF for commercial and military purposes became apparent. Amateur experimentation was then confined only to authorized frequency segments in that range.

Direct Modes (Line-of-sight)

Line-of-sight is the direct propagation of radio waves between antennas that are visible to each other. This is probably the most common of the radio propagation modes at VHF and higher frequencies. Because radio signals can travel through many non-metallic objects, radio can be picked up through walls. This is still line-of-sight propagation. Examples would include propagation between a satellite and a ground antenna or reception of television signals from a local TV transmitter.

Ground plane reflection effects are an important factor in VHF line of sight propagation. The interference between the direct beam line-of-sight and the ground reflected beam often leads to an effective inverse-fourth-power (1/distance^4) law for ground-plane limited radiation.

Ionospheric Modes (Skywave)

Skywave propagation, also referred to as skip, is any of the modes that rely on refraction of radio waves in the ionosphere, which is made up of one or more ionized layers in the upper atmosphere. F2-layer is the most important ionospheric layer for long-distance, multiple-hop HF propagation, though F1, E, and D-layers also play significant roles. The D-layer, when present during sunlight periods, causes significant amount of signal loss, as does the E-layer whose maximum usable frequency can rise to 4 MHz and above and thus block higher frequency signals from reaching the F2-layer. The layers, or more appropriately "regions", are directly affected by the sun on a daily diurnal cycle, a seasonal cycle and the 11-year sunspot cycle and determine the utility of these modes. During solar maxima, or sunspot highs and peaks, the whole HF range up to 30 MHz can be used usually around the clock and F2 propagation up to 50 MHz is observed frequently depending upon daily solar flux 10.7cm radiation values. During solar minima, or minimum sunspot counts down to zero, propagation of frequencies above 15 MHz is generally unavailable.

Although the claim is commonly made that two-way HF propagation along a given path is reciprocal, that is, if the signal from location A reaches location B at a good strength, the signal from location B will be similar at station A because the same path is traversed in both directions. However, the ionosphere is far too complex and constantly changing to support the reciprocity theorem. The path is never exactly the same in both directions. In brief, conditions at the two terminii of a path generally cause dissimilar polarization shifts, dissimilar splits into ordinary rays and extraordinary or *Pedersen rays* which are erratic and impossibly identical or similar due to variations in ionization density, shifting zenith angles, effects of the earth's magnetic DIPOLE contours, antenna radiation patterns, ground conditions and other variables.

Forecasting of skywave modes is of considerable interest to amateur radio operators and commercial marine and aircraft communications, and also to shortwave broadcasters. Real-time propagation can be assessed by listening for transmissions from specific beacon transmitters.

Meteor Scattering

Meteor scattering relies on reflecting radio waves off the intensely ionized columns of air generated by meteors. While this mode is very short duration, often only from a fraction of second to couple of seconds per event, digital Meteor burst communications allows remote stations to communicate to a station that may be hundreds of miles up to over 1,000 miles (1,600 km) away, without the expense required for a satellite link. This mode is most generally useful on VHF frequencies between 30 and 250 MHz.

Auroral Backscatter

Intense columns of Auroral ionization at 100 km altitudes within the auroral oval backscatter radio waves, perhaps most notably on HF and VHF. Backscatter is angle-sensitive—incident ray vs. magnetic field line of the column must be very close to right-angle. Random motions of electrons spiraling around the field lines create a Doppler-spread that broadens the spectra of the emission to more or less noise-like—depending on how high radio frequency is used. The radio-auroras are observed mostly at high latitudes and rarely extend down to middle latitudes. The occurrence of radio-auroras depends on solar activity (flares, coronal holes, CMEs) and annually the events are more numerous during solar cycle maxima. Radio aurora includes the so-called afternoon radio aurora which produces stronger but more distorted signals and after the Harang-minima, the late-night radio aurora (sub-storming phase) returns with variable signal strength and lesser doppler spread. The propagation range for this predominantly back-scatter mode extends up to about 2000 km in east-west plane, but strongest signals are observed most frequently from the north at nearby sites on same latitudes.

Rarely, a strong radio-aurora is followed by Auroral-E, which resembles both propagation types in some ways.

Sporadic-E Propagation

Sporadic E (Es) propagation can be observed on HF and VHF bands. It must not be confused with ordinary HF E-layer propagation. Sporadic-E at mid-latitudes occurs mostly during summer season, from May to August in the northern hemisphere and from November to February in the southern hemisphere. There is no single cause for this mysterious propagation mode. The reflection takes place in a thin sheet of ionisation around 90 km height. The ionisation patches drift westwards at speeds of few hundred km per hour. There is a weak periodicity noted during the season and typically Es is observed on 1 to 3 successive days and remains absent for a few days to reoccur again. Es do not occur during small hours; the events usually begin at dawn, and there is a peak in the afternoon and a second peak in the evening. Es propagation is usually gone by local midnight.

Observation of radio propagation beacons operating around 28.2 MHz, 50 MHz and 70 MHz, indicates that maximum observed frequency (MOF) for Es is found to be lurking around 30 MHz on most days during the summer season, but sometimes MOF may shoot up to 100 MHz or even more in ten minutes to decline slowly during the next few hours. The peak-phase includes oscillation of MOF with periodicity of approximately 5...10 minutes. The propagation range for Es single-hop is typically 1000 to 2000 km, but with multi-hop, double range is observed. The signals are very strong but also with slow deep fading.

Tropospheric Modes

Tropospheric Scattering

At VHF and higher frequencies, small variations (turbulence) in the density of the atmosphere at a height of around 6 miles (10 km) can scatter some of the normally line-of-sight beam of radio frequency energy back toward the ground, allowing over-the-horizon communication between stations as far as 500 miles (800 km) apart. The military developed the White Alice Communications System covering all of Alaska, using this tropospheric scattering principle.

Tropospheric Ducting

Sudden changes in the atmosphere's vertical moisture content and temperature profiles can on random occasions make microwave and UHF & VHF signals propagate hundreds of kilometers up to about

2,000 kilometers (1,300 mi)—and for ducting mode even farther—beyond the normal radio-horizon. The inversion layer is mostly observed over high pressure regions, but there are several tropospheric weather conditions which create these randomly occurring propagation modes. Inversion layer's altitude for non-ducting is typically found between 100 meters (300 ft) to about 1 kilometer (3,000 ft) and for ducting about 500 meters to 3 kilometers (1,600 to 10,000 ft), and the duration of the events are typically from several hours up to several days. Higher frequencies experience the most dramatic increase of signal strengths, while on low-VHF and HF the effect is negligible. Propagation path attenuation may be below free-space loss. Some of the lesser inversion types related to warm ground and cooler air moisture content occur regularly at certain times of the year and time of day. A typical example could be the late summer, early morning tropospheric enhancements that bring in signals from distances up to few hundred kilometers for a couple of hours, until undone by the Sun's warming effect.

Tropospheric Delay

This is a source of error in radio ranging techniques, such as the Global Positioning System (GPS).

Rain Scattering

Rain scattering is purely a microwave propagation mode and is best observed around 10 GHz, but extends down to a few gigahertz—the limit being the size of the scattering particle size vs. wavelength. This mode scatters signals mostly forwards and backwards when using horizontal polarization and side-scattering with vertical polarization. Forward-scattering typically yields propagation ranges of 800 km. Scattering from snowflakes and ice pellets also occurs, but scattering from ice without watery surface is less effective. The most common application for this phenomenon is microwave rain radar, but rain scatter propagation can be a nuisance causing unwanted signals to intermittently propagate where they are not anticipated or desired. Similar reflections may also occur from insects though at lower altitudes and shorter range. Rain also causes attenuation of point-to-point and satellite microwave links. Attenuation values up to 30 dB have been observed on 30 GHz during heavy tropical rain.

Airplane Scattering

Airplane scattering (or most often reflection) is observed on VHF through microwaves and, besides back-scattering, yields momentary propagation up to 500 km even in mountainous terrain. The most common back-scatter applications are air-traffic radar, bistatic forward-scatter guided-missile and airplane-detecting trip-wire radar, and the US space radar.

Lightning Scattering

Lightning scattering has sometimes been observed on VHF and UHF over distances of about 500 km. The hot lightning channel scatters radio-waves for a fraction of a second. The RF noise burst from the lightning makes the initial part of the open channel unusable and the ionization disappears quickly because of recombination at low altitude and high atmospheric pressure. Although the hot lightning channel is briefly observable with microwave radar, no practical use for this mode has been found in communications.

Other Effects

Diffraction

Knife-edge diffraction is the propagation mode where radio waves are bent around sharp edges. For example, this mode is used to send radio signals over a mountain range when a line-of-sight path is not available. However, the angle cannot be too sharp or the signal will not diffract. The diffraction mode requires increased signal strength, so higher power or better antennas will be needed than for an equivalent line-of-sight path.

Diffraction depends on the relationship between the wavelength and the size of the obstacle. In other words, the size of the obstacle in wavelengths. Lower frequencies diffract around large smooth obstacles such as hills more easily. For example, in many cases where VHF (or higher frequency) communication is not possible due to shadowing by a hill, it is still possible to communicate using the upper part of the HF band where the surface wave is of little use.

Diffraction phenomena by small obstacles are also important at high frequencies. Signals for urban cellular telephony tend to be dominated by ground-plane effects as they travel over the rooftops of the urban environment. They then diffract over roof edges into the street, where multipath propagation, absorption and diffraction phenomena dominate.

Absorption

Low-frequency radio waves travel easily through brick and stone and VLF even penetrates sea-water. As the frequency rises, absorption effects become more important. At microwave or higher frequencies, absorption by molecular resonances in the atmosphere (mostly from water, H_2O and oxygen, O_2) is a major factor in radio propagation. For example, in the 58–60 GHz band, there is a major absorption peak which makes this band useless for long-distance use. This phenomenon was first discovered during radar research in World War II. Above about 400 GHz, the Earth's atmosphere blocks most of the spectrum while still passing some - up to UV light, which is blocked by ozone - but visible light and some of the near-infrared is transmitted. Heavy rain and falling snow also affect microwave absorption.

Measuring HF Propagation

HF propagation conditions can be simulated using radio propagation models, such as the Voice of America Coverage Analysis Program, and realtime measurements can be done using chirp transmitters. For radio amateurs the WSPR mode provides maps with real time propagation conditions between a network of transmitters and receivers. Even without special beacons the realtime propagation conditions can be measured: a worldwide network of receivers decodes morse code signals on amateur radio frequencies in realtime and provides sophisticated search functions and propagation maps for every station received.

Practical Effects

The average person can notice the effects of changes in radio propagation in several ways.

In AM broadcasting, the dramatic ionospheric changes that occur overnight in the mediumwave band drive a unique broadcast license scheme, with entirely different transmitter power output levels and directional antenna patterns to cope with skywave propagation at night. Very few stations are allowed to run without modifications during dark hours, typically only those on clear channels in North America. Many stations have no authorization to run at all outside of daylight hours. Otherwise, there would be nothing but interference on the entire broadcast band from dusk until dawn without these modifications.

For FM broadcasting (and the few remaining low-band TV stations), weather is the primary cause for changes in VHF propagation, along with some diurnal changes when the sky is mostly without cloud cover. These changes are most obvious during temperature inversions, such as in the late-night and early-morning hours when it is clear, allowing the ground and the air near it to cool more rapidly. This not only causes dew, frost, or fog, but also causes a slight "drag" on the bottom of the radio waves, bending the signals down such that they can follow the Earth's curvature over the normal radio horizon. The result is typically several stations being heard from another media market — usually a neighboring one, but sometimes ones from a few hundred kilometers away. Ice storms are also the result of inversions, but these normally cause more scattered omnidirection propagation, resulting mainly in interference, often among weather radio stations. In late spring and early summer, a combination of other atmospheric factors can occasionally cause skips that duct high-power signals to places well over 1000km away.

Non-broadcast signals are also affected. Mobile phone signals are in the UHF band, ranging from 700 to over 2600 Megahertz, a range which makes them even more prone to weather-induced propagation changes. In urban (and to some extent suburban) areas with a high population density, this is partly offset by the use of smaller cells, which use lower effective radiated power and beam tilt to reduce interference, and therefore increase frequency reuse and user capacity. However, since this would not be very cost-effective in more rural areas, these cells are larger and so more likely to cause interference over longer distances when propagation conditions allow.

While this is generally transparent to the user thanks to the way that cellular networks handle cell-to-cell handoffs, when cross-border signals are involved, unexepected charges for international roaming may occur despite not having left the country at all. This often occurs between southern San Diego and northern Tijuana at the western end of the U.S./Mexico border, and between eastern Detroit and western Windsor along the U.S./Canada border. Since signals can travel unobstructed over a body of water far larger than the Detroit River, and cool water temperatures also cause inversions in surface air, this "fringe roaming" sometimes occurs across the Great Lakes, and between islands in the Caribbean. Signals can skip from the Dominican Republic to a mountainside in Puerto Rico and vice versa, or between the U.S. and British Virgin Islands, among others. While unintended cross-border roaming is often automatically removed by mobile phone company billing systems, inter-island roaming is typically not.

Radio Frequency

Radio frequency (RF) is any of the electromagnetic wave frequencies that lie in the range extending from around 3 kHz to 300 GHz, which include those frequencies used for communications or radar signals. RF usually refers to electrical rather than mechanical oscillations. However, mechanical RF systems do exist.

Although radio *frequency* is a rate of oscillation, the term “radio frequency” or its abbreviation “RF” are used as a synonym for radio – i.e., to describe the use of wireless communication, as opposed to communication via electric wires. Examples include:

- Radio-frequency identification
- ISO/IEC 14443-2 *Radio frequency power and signal interface*

Special Properties of RF Current

Electric currents that oscillate at radio frequencies have special properties not shared by direct current or alternating current of lower frequencies.

- The energy in an RF current can radiate off a conductor into space as electromagnetic waves (radio waves); this is the basis of radio technology.
- RF current does not penetrate deeply into electrical conductors but tends to flow along their surfaces; this is known as the skin effect. For this reason, when the human body comes in contact with high power RF currents it can cause superficial but serious burns called *RF burns (Note that RF burns result from electrical energy, while Radiation burns result from electromagnetic energy).*
- RF currents applied to the body often do not cause the painful sensation of electric shock as do lower frequency currents. This is because the current changes direction too quickly to trigger depolarization of nerve membranes.
- RF current can easily ionize air, creating a conductive path through it. This property is exploited by “high frequency” units used in electric arc welding, which use currents at higher frequencies than power distribution uses.
- Another property is the ability to appear to flow through paths that contain insulating material, like the dielectric insulator of a capacitor.
- When conducted by an ordinary electric cable, RF current has a tendency to reflect from discontinuities in the cable such as connectors and travel back down the cable toward the source, causing a condition called standing waves. Therefore, RF current must be carried by specialized types of cable called transmission line.

Radio Communication

To receive radio signals an antenna must be used. However, since the antenna will pick up thousands of radio signals at a time, a radio tuner is necessary to *tune into* a particular frequency (or frequency range). This is typically done via a resonator – in its simplest form, a circuit with a capacitor and an inductor form a tuned circuit. The resonator amplifies oscillations within a particular frequency band, while reducing oscillations at other frequencies outside the band. Another method to isolate a particular radio frequency is by oversampling (which gets a wide range of frequencies) and picking out the frequencies of interest, as done in software defined radio.

The distance over which radio communications is useful depends significantly on things other than wavelength, such as transmitter power, receiver quality, type, size, and height of antenna, mode of transmission, noise, and interfering signals. Ground waves, tropospheric scatter and skywaves can all achieve greater ranges than line-of-sight propagation. The study of radio propagation allows estimates of useful range to be made.

Frequency Bands

Frequency	Wavelength	Designation	Abbreviation
3–30 Hz	10^5–10^4 km	Extremely low frequency	ELF
30–300 Hz	10^4–10^3 km	Super low frequency	SLF
300–3000 Hz	10^3–100 km	Ultra low frequency	ULF
3–30 kHz	100–10 km	Very low frequency	VLF
30–300 kHz	10–1 km	Low frequency	LF
300 kHz – 3 MHz	1 km – 100 m	Medium frequency	MF
3–30 MHz	100–10 m	High frequency	HF
30–300 MHz	10–1 m	Very high frequency	VHF
300 MHz – 3 GHz	1 m – 10 cm	Ultra high frequency	UHF
3–30 GHz	10–1 cm	Super high frequency	SHF
30–300 GHz	1 cm – 1 mm	Extremely high frequency	EHF
300 GHz – 3 THz	1 mm – 0.1 mm	Tremendously high frequency	THF

In Medicine

Radio frequency (RF) energy, in the form of radiating waves or electrical currents, has been used in medical treatments for over 75 years, generally for minimally invasive surgeries using radiofrequency ablation including the treatment of sleep apnea. Magnetic resonance imaging (MRI) uses radio frequency waves to generate images of the human body.

Radio frequencies at non-ablation energy levels are sometimes used as a form of cosmetic treatment that can tighten skin, reduce fat (lipolysis), or promote healing.

RF diathermy is a medical treatment that uses RF induced heat as a form of physical or occupational therapy and in surgical procedures. It is commonly used for muscle relaxation. It is also a method of heating tissue electromagnetically for therapeutic purposes in medicine. Diathermy is used in physical therapy and occupational therapy to deliver moderate heat directly to pathologic lesions in the deeper tissues of the body. Surgically, the extreme heat that can be produced by diathermy may be used to destroy neoplasms, warts, and infected tissues, and to cauterize blood vessels to prevent excessive bleeding. The technique is particularly valuable in neurosurgery and surgery of the eye. Diathermy equipment typically operates in the short-wave radio frequency (range 1–100 MHz) or microwave energy (range 434–915 MHz).

Pulsed electromagnetic field therapy (PEMF) is a medical treatment that purportedly helps to heal bone tissue reported in a recent NASA study. This method usually employs electromagnetic radiation of different frequencies - ranging from static magnetic fields, through extremely low frequencies (ELF) to higher radio frequencies (RF) administered in pulses.

Effects on the Human Body

Extremely Low Frequency RF

High-power extremely low frequency RF with electric field levels in the low kV/m range are known to induce perceivable currents within the human body that create an annoying tingling sensation. These currents will typically flow to ground through a body contact surface such as the feet, or arc to ground where the body is well insulated.

Microwaves

Microwave exposure at low-power levels below the Specific absorption rate set by government regulatory bodies are considered harmless non-ionizing radiation and have no effect on the human body. However, levels above the Specific absorption rate set by the U.S. Federal Communications Commission are considered potentially harmful.

Long-term human exposure to high-levels of microwaves is recognized to cause cataracts according to experimental animal studies and epidemiological studies. The mechanism is unclear but may include changes in heat sensitive enzymes that normally protect cell proteins in the lens. Another mechanism that has been advanced is direct damage to the lens from pressure waves induced in the aqueous humor.

High-power exposure to microwave RF is known to create a range of effects from lower to higher power levels, ranging from unpleasant burning sensation on the skin and microwave auditory effect, to extreme pain at the mid-range, to physical burning and blistering of skin and internals at high power levels.

General RF Exposure

The 1999 revision of Canadian Safety Code 6 recommended electric field limits of 100 kV/m for pulsed EMF to prevent air breakdown and spark discharges, mentioning rationale related to auditory effect and energy-induced unconsciousness in rats. The pulsed EMF limit was removed in later revisions, however.

As a Weapon

A heat ray is an RF harassment device that makes use of microwave radio frequencies to create an unpleasant heating effect in the upper layer of the skin. A publicly known heat ray weapon called the Active Denial System was developed by the US military as an experimental weapon to deny the enemy access to an area. A death ray is a weapon that delivers heat ray electromagnetic energy at levels that injure human tissue. The inventor of the death ray, Harry Grindell Matthews, claims to have lost sight in his left eye while developing his death ray weapon based on a primitive microwave magnetron from the 1920s (note that a typical microwave oven induces a tissue damaging cooking effect inside the oven at about 2 kV/m.).

Measurement

Since radio frequency radiation has both an electric and a magnetic component, it is often convenient to express intensity of radiation field in terms of units specific to each component. The unit *volts per meter* (V/m) is used for the electric component, and the unit *amperes per meter* (A/m) is used for the magnetic component. One can speak of an electromagnetic field, and these units are used to provide information about the levels of electric and magnetic field strength at a measurement location.

Another commonly used unit for characterizing an RF electromagnetic field is *power density*. Power density is most accurately used when the point of measurement is far enough away from the RF emitter to be located in what is referred to as the far field zone of the radiation pattern. In closer proximity to the transmitter, i.e., in the "near field" zone, the physical relationships between the electric and magnetic components of the field can be complex, and it is best to use the field strength units discussed above. Power density is measured in terms of power per unit area, for example, milliwatts per square centimeter (mW/cm^2). When speaking of frequencies in the microwave range and higher, power density is usually used to express intensity since exposures that might occur would likely be in the far field zone.

Skywave

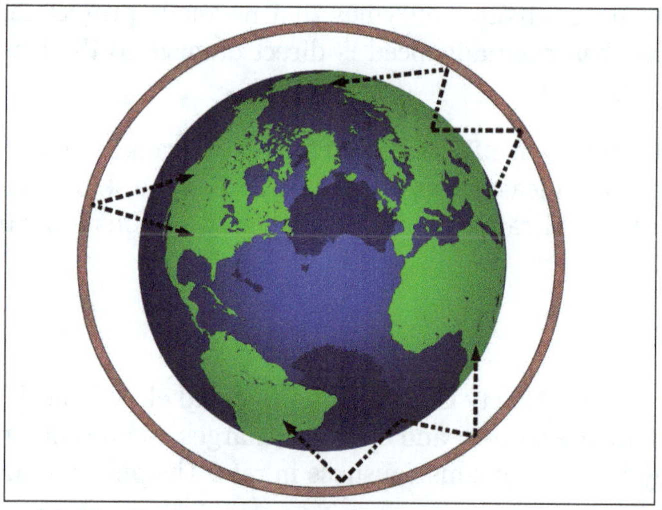

Radio waves (black) reflecting off the ionosphere (red) during skywave propagation.

In radio communication, skywave or skip refers to the propagation of radio waves reflected or refracted back toward Earth from the ionosphere, an electrically charged layer of the upper atmosphere. Since it is not limited by the curvature of the Earth, skywave propagation can be used to communicate beyond the horizon, at intercontinental distances. It is mostly used in the shortwave frequency bands.

As a result of skywave propagation, a signal from a distant AM broadcasting station, a shortwave station, or—during sporadic E propagation conditions (principally during the summer months in both hemispheres)—a low frequency television station can sometimes be received as clearly as local stations. Most long-distance shortwave (high frequency) radio communication—between 3 and 30 MHz—is a result of skywave propagation. Since the early 1920s amateur radio operators (or "hams"), limited to lower transmitter power than broadcast stations, have taken advantage of skywave for long distance (or "DX") communication.

Skywave propagation is distinct from:

- Groundwave propagation, where radio waves travel near Earth's surface without being reflected or refracted by the atmosphere—the dominant propagation mode at lower frequencies,
- Line-of-sight propagation, in which radio waves travel in a straight line, the dominant mode at higher frequencies.

Explanation

The ionosphere is a region of the upper atmosphere, from about 80 km to 1000 km in altitude, where neutral air is ionized by solar photons and cosmic rays. When high frequency signals enter the ionosphere obliquely, they are back-scattered from the ionized layer as scatter waves. If the midlayer ionization is strong enough compared to the signal frequency, a scatter wave can exit the bottom of the layer earthwards as if reflected from a mirror. Earth's surface (ground or water) then diffusely reflects the incoming wave back towards the ionosphere. Consequently, like a rock "skipping" across water, the signal may effectively "bounce" or "skip" between the earth and ionosphere two or more times (multihop propagation). Since at shallow incidence losses remain quite small, signals of only a few watts can sometimes be received many thousands of miles away as a result. This is what enables shortwave broadcasts to travel all over the world.

If the ionization is not great enough, the scatter wave is initially deflected downwards, and subsequently upwards (above the layer peak) such that it exits the top of the layer slightly displaced. Sky wave propagation occurs in the waveguide formed by the ground and ionosphere, each serving as reflectors. With a single "hop," path distances up to 3500 km may be reached. Transatlantic connections are mostly obtained with two or three hops.

The layer of ionospheric plasma with equal ionization (the reflective surface) is not fixed, but undulates like the surface of the ocean. Varying reflection efficiency from this changing surface can cause the reflected signal strength to change, causing "*fading*" in shortwave broadcasts.

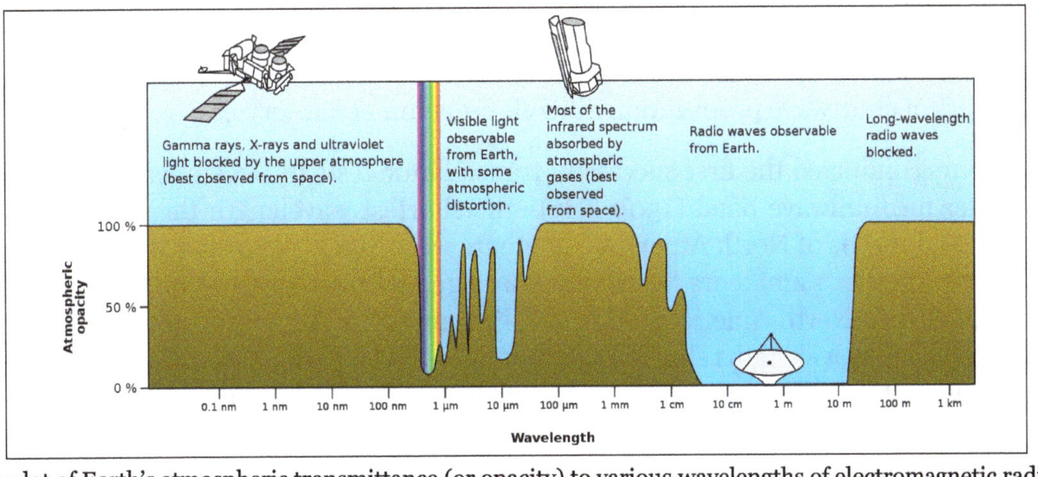

Rough plot of Earth's atmospheric transmittance (or opacity) to various wavelengths of electromagnetic radiation, including radio waves.

Depending on the transmitting antenna, signals below approximately 10 MHz during the day and 5 MHz at night, entering the ionosphere at a steep angle (near-vertical incidence) may be back-scattered down to Earth within a short range. Alternatively, signals beamed close to the horizon enter the ionosphere at a shallow angle and return to Earth over medium to long distances.

Other Considerations

VHF signals with frequencies above about 30 MHz usually penetrate the ionosphere and are not returned to the Earth's surface. E-skip is a notable exception, where VHF signals including FM broadcast and VHF TV signals are frequently reflected to the Earth during late Spring and early Summer. E-skip rarely affects UHF frequencies, except for very rare occurrences below 500 MHz.

Frequencies below approximately 10 MHz (wavelengths longer than 30 meters), including broadcasts in the mediumwave and shortwave bands (and to some extent longwave), propagate most efficiently by skywave at night. Frequencies above 10 MHz (wavelengths shorter than 30 meters) typically propagate most efficiently during the day. Frequencies lower than 3 kHz have a wavelength longer than the distance between the Earth and the ionosphere. The maximum usable frequency for skywave propagation is strongly influenced by sunspot number.

Skywave propagation is usually degraded—sometimes seriously—during geomagnetic storms. Skywave propagation on the sunlit side of the Earth can be entirely disrupted during sudden ionospheric disturbances.

Because the lower-altitude layers (the E-layer in particular) of the ionosphere largely disappear at night, the refractive layer of the ionosphere is much higher above the surface of the Earth at night. This leads to an increase in the "skip" or "hop" distance of the skywave at night.

History

Discovery of Skywave Propagation

Amateur radio operators are credited with the discovery of skywave propagation on the shortwave bands. Early long-distance services used surface wave propagation at very low frequencies, which are attenuated along the path. Longer distances and higher frequencies using this method meant more signal attenuation. This, and the difficulties of generating and detecting higher frequencies, made discovery of shortwave propagation difficult for commercial services.

Radio amateurs conducted the first successful transatlantic tests in December 1921, operating in the 200 meter mediumwave band (1500 kHz)—the shortest wavelength then available to amateurs. In 1922 hundreds of North American amateurs were heard in Europe at 200 meters and at least 30 North American amateurs heard amateur signals from Europe. The first two-way communications between North American and Hawaiian amateurs began in 1922 at 200 meters. Although operation on wavelengths shorter than 200 meters was technically illegal (but tolerated as the authorities mistakenly believed at first that such frequencies were useless for commercial or military use), amateurs began to experiment with those wavelengths using newly available vacuum tubes shortly after World War I.

Extreme interference at the upper edge of the 150-200 meter band—the official wavelengths allocated to amateurs by the Second National Radio Conference in 1923—forced amateurs to shift to shorter and shorter wavelengths; however, amateurs were limited by regulation to wavelengths longer than 150 meters (2 MHz). A few fortunate amateurs who obtained special permission for experimental communications below 150 meters completed hundreds of long distance two way contacts on 100 meters (3 MHz) in 1923 including the first transatlantic two way contacts in November 1923, on 110 meters (2.72 MHz).

By 1924 many additional specially licensed amateurs were routinely making transoceanic contacts at distances of 6000 miles (~9600 km) and more. On 21 September several amateurs in California completed two way contacts with an amateur in New Zealand. On 19 October amateurs in New Zealand and England completed a 90-minute two-way contact nearly halfway around the world. On October 10, the Third National Radio Conference made three shortwave bands available to U.S. amateurs at 80 meters (3.75 MHz), 40 meters (7 MHz) and 20 meters (14 MHz). These were allocated worldwide, while the 10-meter band (28 MHz) was created by the Washington International Radiotelegraph Conference on 25 November 1927. The 15-meter band (21 MHz) was opened to amateurs in the United States on 1 May 1952.

Marconi

In June and July 1923, Guglielmo Marconi's transmissions were completed during nights on 97 meters from Poldhu Wireless Station, Cornwall, to his yacht Ellette in the Cape Verde Islands. In September 1924, Marconi transmitted during daytime and nighttime on 32 meters from Poldhu to his yacht in Beirut. Marconi, in July 1924, entered into contracts with the British General Post Office (GPO) to install high speed shortwave telegraphy circuits from London to Australia, India, South Africa and Canada as the main element of the Imperial Wireless Chain. The UK-to-Canada shortwave "Beam Wireless Service" went into commercial operation on 25 October 1926. Beam Wireless Services from the UK to Australia, South Africa and India went into service in 1927.

Far more spectrum is available for long distance communication in the shortwave bands than in the long wave bands; and shortwave transmitters, receivers and antennas were orders of magnitude less expensive than the multi-hundred kilowatt transmitters and monstrous antennas needed for long wave.

Shortwave communications began to grow rapidly in the 1920s, similar to the internet in the late 20th century. By 1928, more than half of long distance communications had moved from transoceanic cables and long wave wireless services to shortwave "skip" transmission and the overall volume of transoceanic shortwave communications had vastly increased. Shortwave also ended the need for multimillion-dollar investments in new transoceanic telegraph cables and massive long wave wireless stations, although some existing transoceanic telegraph cables and commercial long wave communications stations remained in use until the 1960s.

The cable companies began to lose large sums of money in 1927, and a serious financial crisis threatened the viability of cable companies that were vital to strategic British interests. The British government convened the Imperial Wireless and Cable Conference in 1928 "to examine the situation that had arisen as a result of the competition of Beam Wireless with the Cable Services". It recommended and received Government approval for all overseas cable and wireless resources of the Empire to be merged into one system controlled by a newly formed company in 1929, Imperial and International Communications Ltd. The name of the company was changed to Cable and Wireless Ltd. in 1934.

References

- Melvin A. Shiffman, Sid J. Mirrafati, Samuel M. Lam and Chelso G. Cueteaux (2007). Simplified Facial Rejuvenation. Springer. p. 157. ISBN 978-3-540-71096-7
- Clint Smith, Curt Gervelis (2003). Wireless Network Performance Handbook. McGraw-Hill Professional. ISBN 0-07-140655-7
- Free radio: electronic civil disobedience by Lawrence C. Soley. Published by Westview Press, 1998. ISBN 0-8133-9064-8, ISBN 978-0-8133-9064-2
- Rebel Radio: The Full Story of British Pirate Radio by John Hind, Stephen Mosco. Published by Pluto Press, 1985. ISBN 0-7453-0055-3, ISBN 978-0-7453-0055-9
- Ruey J. Sung & Michael R. Lauer (2000). Fundamental approaches to the management of cardiac arrhythmias. Springer. p. 153. ISBN 978-0-7923-6559-4
- Nebeker, Frederik (2009). Dawn of the Electronic Age: Electrical Technologies in the Shaping of the Modern World, 1914 to 1945. John Wiley & Sons. pp. 14–15. ISBN 0470409746

6

Security for Wireless Networks

The prevention of unauthorized damage to computers which are connected to wireless networks is known as wireless security. The process of monitoring unauthorized access points is known as intrusion detection, and the device that performs intrusion detection and takes countermeasures is known as the wireless intrusion prevention system. This chapter discusses in detail the theories and concepts related to wireless security.

Wireless Security

Wireless security is the prevention of unauthorized access or damage to computers using wireless networks. The most common types of wireless security are Wired Equivalent Privacy (WEP) and Wi-Fi Protected Access (WPA). WEP is a notoriously weak security standard. The password it uses can often be cracked in a few minutes with a basic laptop computer and widely available software tools. WEP is an old IEEE 802.11 standard from 1999, which was outdated in 2003 by WPA, or Wi-Fi Protected Access. WPA was a quick alternative to improve security over WEP. The current standard is WPA2; some hardware cannot support WPA2 without firmware upgrade or replacement. WPA2 uses an encryption device that encrypts the network with a 256-bit key; the longer key length improves security over WEP.

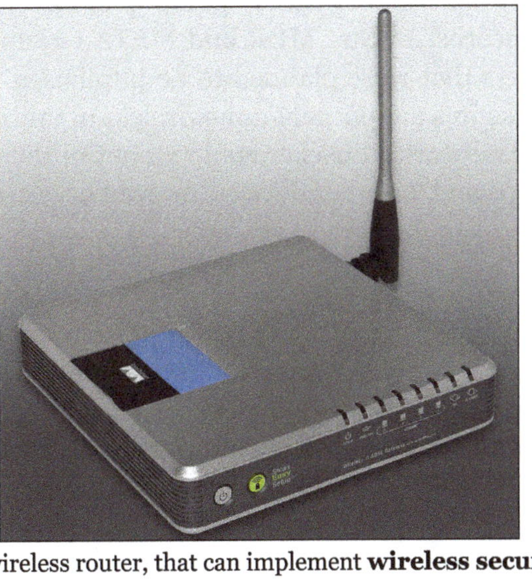

An example wireless router, that can implement **wireless security** features.

Many laptop computers have wireless cards pre-installed. The ability to enter a network while mobile has great benefits. However, wireless networking is prone to some security issues. Hackers have found wireless networks relatively easy to break into, and even use wireless technology to hack into wired networks. As a result, it is very important that enterprises define effective wireless security policies that guard against unauthorized access to important resources. Wireless Intrusion Prevention Systems (WIPS) or Wireless Intrusion Detection Systems (WIDS) are commonly used to enforce wireless security policies.

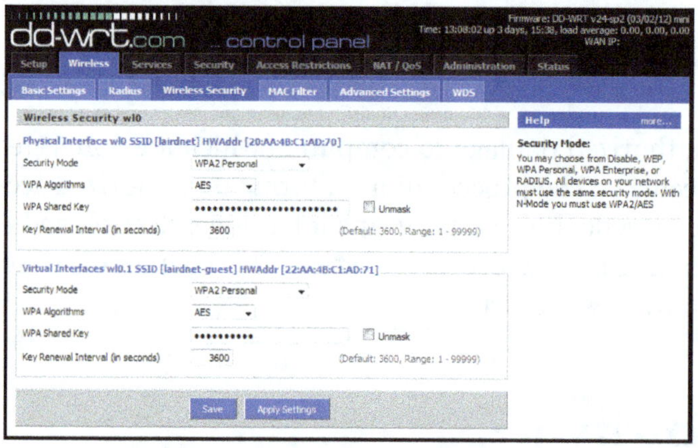

Security settings panel for a DD-WRT router.

The risks to users of wireless technology have increased as the service has become more popular. There were relatively few dangers when wireless technology was first introduced. Hackers had not yet had time to latch on to the new technology, and wireless networks were not commonly found in the work place. However, there are many security risks associated with the current wireless protocols and encryption methods, and in the carelessness and ignorance that exists at the user and corporate IT level. Hacking methods have become much more sophisticated and innovative with wireless access. Hacking has also become much easier and more accessible with easy-to-use Windows- or Linux-based tools being made available on the web at no charge.

Some organizations that have no wireless access points installed do not feel that they need to address wireless security concerns. In-Stat MDR and META Group have estimated that 95% of all corporate laptop computers that were planned to be purchased in 2005 were equipped with wireless cards. Issues can arise in a supposedly non-wireless organization when a wireless laptop is plugged into the corporate network. A hacker could sit out in the parking lot and gather information from it through laptops and/or other devices, or even break in through this wireless card–equipped laptop and gain access to the wired network.

Background

Anyone within the geographical network range of an open, unencrypted wireless network can "sniff", or capture and record, the traffic, gain unauthorized access to internal network resources as well as to the internet, and then use the information and resources to perform disruptive or illegal acts. Such security breaches have become important concerns for both enterprise and home networks.

If router security is not activated or if the owner deactivates it for convenience, it creates a free hotspot. Since most 21st-century laptop PCs have wireless networking built in, they don't need a third-party adapter such as a PCMCIA Card or USB dongle. Built-in wireless networking might be enabled by default, without the owner realizing it, thus broadcasting the laptop's accessibility to any computer nearby.

Modern operating systems such as Linux, macOS, or Microsoft Windows make it fairly easy to set up a PC as a wireless LAN "base station" using Internet Connection Sharing, thus allowing all the PCs in the home to access the Internet through the "base" PC. However, lack of knowledge among users about the security issues inherent in setting up such systems often may allow others nearby access to the connection. Such "piggybacking" is usually achieved without the wireless network operator's knowledge; it may even be without the knowledge of the intruding user if their computer automatically selects a nearby unsecured wireless network to use as an access point.

The Threat Situation

Wireless security is just an aspect of computer security; however, organizations may be particularly vulnerable to security breaches caused by rogue access points.

If an employee (trusted entity) brings in a wireless router and plugs it into an unsecured switchport, the entire network can be exposed to anyone within range of the signals. Similarly, if an employee adds a wireless interface to a networked computer using an open USB port, they may create a breach in network security that would allow access to confidential materials. However, there are effective countermeasures (like disabling open switchports during switch configuration and VLAN configuration to limit network access) that are available to protect both the network and the information it contains, but such countermeasures must be applied uniformly to all network devices.

Threats and Vulnerabilites in an Industrial (M2M) Context

Due to its availability and low cost, the use of wireless communication technologies increases in domains beyond the originally intended usage areas, e.g. M2M communication in industrial applications. Such industrial applications often have specific security requirements. Hence, it is important to understand the characteristics of such applications and evaluate the vulnerabilities bearing the highest risk in this context. Evaluation of these vulnerabilities and the resulting vulnerability catalogs in an industrial context when considering WLAN, NFC and ZigBee are available.

The Mobility Advantage

Wireless networks are very common, both for organizations and individuals. Many laptop computers have wireless cards pre-installed. The ability to enter a network while mobile has great benefits. However, wireless networking is prone to some security issues. Hackers have found wireless networks relatively easy to break into, and even use wireless technology to hack into wired networks. As a result, it is very important that enterprises define effective wireless security policies that guard against unauthorized access to important resources. Wireless Intrusion Prevention Systems (WIPS) or Wireless Intrusion Detection Systems (WIDS) are commonly used to enforce wireless security policies.

The Air Interface and Link Corruption Risk

There were relatively few dangers when wireless technology was first introduced, as the effort to maintain the communication was high and the effort to intrude is always higher. The variety of risks to users of wireless technology have increased as the service has become more popular and the technology more commonly available. Today there are a great number of security risks associated with the current wireless protocols and encryption methods, as carelessness and ignorance exists at the user and corporate IT level. Hacking methods have become much more sophisticated and innovative with wireless.

Modes of Unauthorized Access

The modes of unauthorised access to links, to functions and to data is as variable as the respective entities make use of program code. There does not exist a full scope model of such threat. To some extent the prevention relies on known modes and methods of attack and relevant methods for suppression of the applied methods. However, each new mode of operation will create new options of threatening. Hence prevention requires a steady drive for improvement. The described modes of attack are just a snapshot of typical methods and scenarios where to apply.

Accidental Association

Violation of the security perimeter of a corporate network can come from a number of different methods and intents. One of these methods is referred to as "accidental association". When a user turns on a computer and it latches on to a wireless access point from a neighboring company's overlapping network, the user may not even know that this has occurred. However, it is a security breach in that proprietary company information is exposed and now there could exist a link from one company to the other. This is especially true if the laptop is also hooked to a wired network.

Accidental association is a case of wireless vulnerability called as "mis-association". Mis-association can be accidental, deliberate (for example, done to bypass corporate firewall) or it can result from deliberate attempts on wireless clients to lure them into connecting to attacker's APs.

Malicious Association

"Malicious associations" are when wireless devices can be actively made by attackers to connect to a company network through their laptop instead of a company access point (AP). These types of laptops are known as "soft APs" and are created when a cyber criminal runs some software that makes his/her wireless network card look like a legitimate access point. Once the thief has gained access, he/she can steal passwords, launch attacks on the wired network, or plant trojans. Since wireless networks operate at the Layer 2 level, Layer 3 protections such as network authentication and virtual private networks (VPNs) offer no barrier. Wireless 802.1x authentications do help with some protection but are still vulnerable to hacking. The idea behind this type of attack may not be to break into a VPN or other security measures. Most likely the criminal is just trying to take over the client at the Layer 2 level.

Ad hoc Networks

Ad hoc networks can pose a security threat. Ad hoc networks are defined as [peer to peer] networks between wireless computers that do not have an access point in between them. While these types of networks usually have little protection, encryption methods can be used to provide security.

The security hole provided by Ad hoc networking is not the Ad hoc network itself but the bridge it provides into other networks, usually in the corporate environment, and the unfortunate default settings in most versions of Microsoft Windows to have this feature turned on unless explicitly disabled. Thus the user may not even know they have an unsecured Ad hoc network in operation on their computer. If they are also using a wired or wireless infrastructure network at the same time, they are providing a bridge to the secured organizational network through the unsecured Ad hoc connection. Bridging is in two forms. A direct bridge, which requires the user actually configure a bridge between the two connections and is thus unlikely to be initiated unless explicitly desired, and an indirect bridge which is the shared resources on the user computer. The indirect bridge may expose private data that is shared from the user's computer to LAN connections, such as shared folders or private Network Attached Storage, making no distinction between authenticated or private connections and unauthenticated Ad-Hoc networks. This presents no threats not already familiar to open/public or unsecured wifi access points, but firewall rules may be circumvented in the case of poorly configured operating systems or local settings.

Non-traditional Networks

Non-traditional networks such as personal network Bluetooth devices are not safe from hacking and should be regarded as a security risk. Even barcode readers, handheld PDAs, and wireless printers and copiers should be secured. These non-traditional networks can be easily overlooked by IT personnel who have narrowly focused on laptops and access points.

Identity Theft (MAC Spoofing)

Identity theft (or MAC spoofing) occurs when a hacker is able to listen in on network traffic and identify the MAC address of a computer with network privileges. Most wireless systems allow some kind of MAC filtering to allow only authorized computers with specific MAC IDs to gain access and utilize the network. However, programs exist that have network "sniffing" capabilities. Combine these programs with other software that allow a computer to pretend it has any MAC address that the hacker desires, and the hacker can easily get around that hurdle.

MAC filtering is effective only for small residential (SOHO) networks, since it provides protection only when the wireless device is "off the air". Any 802.11 device "on the air" freely transmits its unencrypted MAC address in its 802.11 headers, and it requires no special equipment or software to detect it. Anyone with an 802.11 receiver (laptop and wireless adapter) and a freeware wireless packet analyzer can obtain the MAC address of any transmitting 802.11 within range. In an organizational environment, where most wireless devices are "on the air" throughout the active working shift, MAC filtering provides only a false sense of security since it prevents only "casual" or unintended connections to the organizational infrastructure and does nothing to prevent a directed attack.

Man-in-the-middle Attacks

A man-in-the-middle attacker entices computers to log in to a computer which is set up as a soft AP (Access Point). Once this is done, the hacker connects to a real access point through another wireless card offering a steady flow of traffic through the transparent hacking computer to the real network. The hacker can then sniff the traffic. One type of man-in-the-middle attack relies on security faults in challenge and handshake protocols to execute a "de-authentication attack". This attack forces AP-connected computers to drop their connections and reconnect with the hacker's soft AP (disconnects the user from the modem so they have to connect again using their password which one can extract from the recording of the event). Man-in-the-middle attacks are enhanced by software such as LANjack and AirJack which automate multiple steps of the process, meaning what once required some skill can now be done by script kiddies. Hotspots are particularly vulnerable to any attack since there is little to no security on these networks.

Denial of Service

A Denial-of-Service attack (DoS) occurs when an attacker continually bombards a targeted AP (Access Point) or network with bogus requests, premature successful connection messages, failure messages, and/or other commands. These cause legitimate users to not be able to get on the network and may even cause the network to crash. These attacks rely on the abuse of protocols such as the Extensible Authentication Protocol (EAP).

The DoS attack in itself does little to expose organizational data to a malicious attacker, since the interruption of the network prevents the flow of data and actually indirectly protects data by preventing it from being transmitted. The usual reason for performing a DoS attack is to observe the recovery of the wireless network, during which all of the initial handshake codes are re-transmitted by all devices, providing an opportunity for the malicious attacker to record these codes and use various cracking tools to analyze security weaknesses and exploit them to gain unauthorized access to the system. This works best on weakly encrypted systems such as WEP, where there are a number of tools available which can launch a dictionary style attack of "possibly accepted" security keys based on the "model" security key captured during the network recovery.

Network Injection

In a network injection attack, a hacker can make use of access points that are exposed to non-filtered network traffic, specifically broadcasting network traffic such as "Spanning Tree" (802.1D), OSPF, RIP, and HSRP. The hacker injects bogus networking re-configuration commands that affect routers, switches, and intelligent hubs. A whole network can be brought down in this manner and require rebooting or even reprogramming of all intelligent networking devices.

Caffe Latte Attack

The Caffe Latte attack is another way to defeat WEP. It is not necessary for the attacker to be in the area of the network using this exploit. By using a process that targets the Windows wireless stack, it is possible to obtain the WEP key from a remote client. By sending a flood of encrypted ARP requests, the assailant takes advantage of the shared key authentication and the message modification flaws in 802.11 WEP. The attacker uses the ARP responses to obtain the WEP key in less than 6 minutes.

Wireless Intrusion Prevention Concepts

There are three principal ways to secure a wireless network.

- For closed networks (like home users and organizations) the most common way is to configure access restrictions in the access points. Those restrictions may include encryption and checks on MAC address. Wireless Intrusion Prevention Systems can be used to provide wireless LAN security in this network model.

- For commercial providers, hotspots, and large organizations, the preferred solution is often to have an open and unencrypted, but completely isolated wireless network. The users will at first have no access to the Internet nor to any local network resources. Commercial providers usually forward all web traffic to a captive portal which provides for payment and/or authorization. Another solution is to require the users to connect securely to a privileged network using VPN.
- Wireless networks are less secure than wired ones; in many offices intruders can easily visit and hook up their own computer to the wired network without problems, gaining access to the network, and it is also often possible for remote intruders to gain access to the network through backdoors like Back Orifice. One general solution may be end-to-end encryption, with independent authentication on all resources that shouldn't be available to the public.

There is no ready designed system to prevent from fraudulent usage of wireless communication or to protect data and functions with wirelessly communicating computers and other entities. However, there is a system of qualifying the taken measures as a whole according to a common understanding what shall be seen as state of the art. The system of qualifying is an international consensus as specified in ISO/IEC 15408.

A Wireless Intrusion Prevention System

A Wireless Intrusion Prevention System (WIPS) is a concept for the most robust way to counteract wireless security risks. However such WIPS does not exist as a ready designed solution to implement as a software package. A WIPS is typically implemented as an overlay to an existing Wireless LAN infrastructure, although it may be deployed standalone to enforce no-wireless policies within an organization. WIPS is considered so important to wireless security that in July 2009, the Payment Card Industry Security Standards Council published wireless guidelines for PCI DSS recommending the use of WIPS to automate wireless scanning and protection for large organizations.

Security Measures

There are a range of wireless security measures, of varying effectiveness and practicality.

SSID Hiding

A simple but ineffective method to attempt to secure a wireless network is to hide the SSID (Service Set Identifier). This provides very little protection against anything but the most casual intrusion efforts.

MAC ID Filtering

One of the simplest techniques is to only allow access from known, pre-approved MAC addresses. Most wireless access points contain some type of MAC ID filtering. However, an attacker can simply sniff the MAC address of an authorized client and spoof this address.

Static IP Addressing

Typical wireless access points provide IP addresses to clients via DHCP. Requiring clients to set their own addresses makes it more difficult for a casual or unsophisticated intruder to log onto the network, but provides little protection against a sophisticated attacker.

802.11 Security

IEEE 802.1X is the IEEE Standard authentication mechanisms to devices wishing to attach to a Wireless LAN.

Regular WEP

The Wired Equivalent Privacy (WEP) encryption standard was the original encryption standard for wireless, but since 2004 with the ratification WPA2 the IEEE has declared it "deprecated", and while often supported, it is seldom or never the default on modern equipment.

Concerns were raised about its security as early as 2001, dramatically demonstrated in 2005 by the FBI, yet in 2007 T.J. Maxx admitted a massive security breach due in part to a reliance on WEP and the Payment Card Industry took until 2008 to prohibit its use - and even then allowed existing use to continue until June 2010.

WPAv1

The Wi-Fi Protected Access (WPA and WPA2) security protocols were later created to address the problems with WEP. If a weak password, such as a dictionary word or short character string is used, WPA and WPA2 can be cracked. Using a long enough random password (e.g. 14 random letters) or passphrase (e.g. 5 randomly chosen words) makes pre-shared key WPA virtually uncrackable. The second generation of the WPA security protocol (WPA2) is based on the final IEEE 802.11i amendment to the 802.11 standard and is eligible for FIPS 140-2 compliance. With all those encryption schemes, any client in the network that knows the keys can read all the traffic.

Wi-Fi Protected Access (WPA) is a software/firmware improvement over WEP. All regular WLAN-equipment that worked with WEP are able to be simply upgraded and no new equipment needs to be bought. WPA is a trimmed-down version of the 802.11i security standard that was developed by the IEEE 802.11 to replace WEP. The TKIP encryption algorithm was developed for WPA to provide improvements to WEP that could be fielded as firmware upgrades to existing 802.11 devices. The WPA profile also provides optional support for the AES-CCMP algorithm that is the preferred algorithm in 802.11i and WPA2.

WPA Enterprise provides RADIUS based authentication using 802.1x. WPA Personal uses a pre-shared Shared Key (PSK) to establish the security using an 8 to 63 character passphrase. The PSK may also be entered as a 64 character hexadecimal string. Weak PSK passphrases can be broken using off-line dictionary attacks by capturing the messages in the four-way exchange when the client reconnects after being deauthenticated. Wireless suites such as aircrack-ng can crack a weak passphrase in less than a minute. Other WEP/WPA crackers are AirSnort and Auditor Security Collection. Still, WPA Personal is secure when used with 'good' passphrases or a full 64-character hexadecimal key.

There was information, however, that Erik Tews (the man who created the fragmentation attack against WEP) was going to reveal a way of breaking the WPA TKIP implementation at Tokyo's PacSec security conference in November 2008, cracking the encryption on a packet in between

12–15 minutes. Still, the announcement of this 'crack' was somewhat overblown by the media, because as of August, 2009, the best attack on WPA (the Beck-Tews attack) is only partially successful in that it only works on short data packets, it cannot decipher the WPA key, and it requires very specific WPA implementations in order to work.

Additions to WPAv1

In addition to WPAv1, TKIP, WIDS and EAP may be added alongside. Also, VPN-networks (non-continuous secure network connections) may be set up under the 802.11-standard. VPN implementations include PPTP, L2TP, IPsec and SSH. However, this extra layer of security may also be cracked with tools such as Anger, Deceit and Ettercap for PPTP; and ike-scan, IKEProbe, ipsectrace, and IKEcrack for IPsec-connections.

TKIP

This stands for Temporal Key Integrity Protocol and the acronym is pronounced as tee-kip. This is part of the IEEE 802.11i standard. TKIP implements per-packet key mixing with a re-keying system and also provides a message integrity check. These avoid the problems of WEP.

EAP

The WPA-improvement over the IEEE 802.1X standard already improved the authentication and authorization for access of wireless and wired LANs. In addition to this, extra measures such as the Extensible Authentication Protocol (EAP) have initiated an even greater amount of security. This, as EAP uses a central authentication server. Unfortunately, during 2002 a Maryland professor discovered some shortcomings. Over the next few years these shortcomings were addressed with the use of TLS and other enhancements. This new version of EAP is now called Extended EAP and is available in several versions; these include: EAP-MD5, PEAPv0, PEAPv1, EAP-MSCHAPv2, LEAP, EAP-FAST, EAP-TLS, EAP-TTLS, MSCHAPv2, and EAP-SIM.

EAP-versions

EAP-versions include LEAP, PEAP and other EAP's.

LEAP

This stands for the Lightweight Extensible Authentication Protocol. This protocol is based on 802.1X and helps minimize the original security flaws by using WEP and a sophisticated key management system. This EAP-version is safer than EAP-MD5. This also uses MAC address authentication. LEAP is not secure; THC-LeapCracker can be used to break Cisco's version of LEAP and be used against computers connected to an access point in the form of a dictionary attack. Anwrap and asleap finally are other crackers capable of breaking LEAP.

PEAP

This stands for Protected Extensible Authentication Protocol. This protocol allows for a secure transport of data, passwords, and encryption keys without the need of a certificate server. This was developed by Cisco, Microsoft, and RSA Security.

Other EAPs There are other types of Extensible Authentication Protocol implementations that are based on the EAP framework. The framework that was established supports existing EAP types as well as future authentication methods. EAP-TLS offers very good protection because of its mutual authentication. Both the client and the network are authenticated using certificates and per-session WEP keys. EAP-FAST also offers good protection. EAP-TTLS is another alternative made by Certicom and Funk Software. It is more convenient as one does not need to distribute certificates to users, yet offers slightly less protection than EAP-TLS.

Restricted Access Networks

Solutions include a newer system for authentication, IEEE 802.1x, that promises to enhance security on both wired and wireless networks. Wireless access points that incorporate technologies like these often also have routers built in, thus becoming wireless gateways.

End-to-end Encryption

One can argue that both layer 2 and layer 3 encryption methods are not good enough for protecting valuable data like passwords and personal emails. Those technologies add encryption only to parts of the communication path, still allowing people to spy on the traffic if they have gained access to the wired network somehow. The solution may be encryption and authorization in the application layer, using technologies like SSL, SSH, GnuPG, PGP and similar.

The disadvantage with the end-to-end method is, it may fail to cover all traffic. With encryption on the router level or VPN, a single switch encrypts all traffic, even UDP and DNS lookups. With end-to-end encryption on the other hand, each service to be secured must have its encryption "turned on", and often every connection must also be "turned on" separately. For sending emails, every recipient must support the encryption method, and must exchange keys correctly. For Web, not all web sites offer https, and even if they do, the browser sends out IP addresses in clear text.

The most prized resource is often access to Internet. An office LAN owner seeking to restrict such access will face the nontrivial enforcement task of having each user authenticate themselves for the router.

802.11i Security

The newest and most rigorous security to implement into WLAN's today is the 802.11i RSN-standard. This full-fledged 802.11i standard (which uses WPAv2) however does require the newest hardware (unlike WPAv1), thus potentially requiring the purchase of new equipment. This new hardware required may be either AES-WRAP (an early version of 802.11i) or the newer and better AES-CCMP-equipment. One should make sure one needs WRAP or CCMP-equipment, as the 2 hardware standards are not compatible.

WPAv2

WPA2 is a WiFi Alliance branded version of the final 802.11i standard. The primary enhancement over WPA is the inclusion of the AES-CCMP algorithm as a mandatory feature. Both WPA and WPA2 support EAP authentication methods using RADIUS servers and preshared key (PSK).

The number of WPA and WPA2 networks are increasing, while the number of WEP networks are decreasing, because of the security vulnerabilities in WEP.

WPA2 has been found to have at least one security vulnerability, nicknamed Hole196. The vulnerability uses the WPA2 Group Temporal Key (GTK), which is a shared key among all users of the same BSSID, to launch attacks on other users of the same BSSID. It is named after page 196 of the IEEE 802.11i specification, where the vulnerability is discussed. In order for this exploit to be performed, the GTK must be known by the attacker.

Additions to WPAv2

Unlike 802.1X, 802.11i already has most other additional security-services such as TKIP. Just as with WPAv1, WPAv2 may work in cooperation with EAP and a WIDS.

WAPI

This stands for WLAN Authentication and Privacy Infrastructure. This is a wireless security standard defined by the Chinese government.

Smart cards, USB Tokens, and Software Tokens

This is a very strong form of security. When combined with some server software, the hardware or software card or token will use its internal identity code combined with a user entered PIN to create a powerful algorithm that will very frequently generate a new encryption code. The server will be time synced to the card or token. This is a very secure way to conduct wireless transmissions. Companies in this area make USB tokens, software tokens, and smart cards. They even make hardware versions that double as an employee picture badge. Currently the safest security measures are the smart cards / USB tokens. However, these are expensive. The next safest methods are WPA2 or WPA with a RADIUS server. Any one of the three will provide a good base foundation for security. The third item on the list is to educate both employees and contractors on security risks and personal preventive measures. It is also IT's task to keep the company workers' knowledge base up-to-date on any new dangers that they should be cautious about. If the employees are educated, there will be a much lower chance that anyone will accidentally cause a breach in security by not locking down their laptop or bring in a wide open home access point to extend their mobile range. Employees need to be made aware that company laptop security extends to outside of their site walls as well. This includes places such as coffee houses where workers can be at their most vulnerable. The last item on the list deals with 24/7 active defense measures to ensure that the company network is secure and compliant. This can take the form of regularly looking at access point, server, and firewall logs to try to detect any unusual activity. For instance, if any large files went through an access point in the early hours of the morning, a serious investigation into the incident would be called for. There are a number of software and hardware devices that can be used to supplement the usual logs and usual other safety measures.

RF Shielding

It's practical in some cases to apply specialized wall paint and window film to a room or building to significantly attenuate wireless signals, which keeps the signals from propagating outside a facility. This can significantly improve wireless security because it's difficult for hackers to receive the signals beyond the controlled area of an enterprise, such as within parking lots.

Denial of Service Defense

Most DoS attacks are easy to detect. However, a lot of them are difficult to stop even after detection. Here are three of the most common ways to stop a DoS attack.

- Black Holing
- Validating the Handshake
- Rate Limiting

Black Holing is one possible way of stopping a DoS attack. This is a situation where we drop all IP packets from an attacker. This is not a very good long-term strategy because attackers can change their source address very quickly.

This may have negative effects if done automatically. An attacker could knowingly spoof attack packets with the IP address of a corporate partner. Automated defenses could block legitimate traffic from that partner and cause additional problems.

Validating the Handshake involves creating false opens, and not setting aside resources until the sender acknowledges. Some firewalls address SYN floods by pre-validating the TCP handshake. This is done by creating false opens. Whenever a SYN segment arrives, the firewall sends back a SYN/ACK segment, without passing the SYN segment on to the target server.

Only when the firewall gets back an ACK, which would happen only in a legitimate connection, would the firewall send the original SYN segment on to the server for which it was originally intended. The firewall doesn't set aside resources for a connection when a SYN segment arrives, so handling a large number of false SYN segments is only a small burden.

Rate limiting can be used to reduce a certain type of traffic down to an amount the can be reasonably dealt with. Broadcasting to the internal network could still be used, but only at a limited rate for example. This is for more subtle DoS attacks. This is good if an attack is aimed at a single server because it keeps transmission lines at least partially open for other communication.

Rate limiting frustrates both the attacker, and the legitimate users. This helps but does not fully solve the problem. Once DoS traffic clogs the access line going to the internet, there is nothing a border firewall can do to help the situation. Most DoS attacks are problems of the community which can only be stopped with the help of ISP's and organizations whose computers are taken over as bots and used to attack other firms.

Mobile Devices

With increasing number of mobile devices with 802.1x interfaces, security of such mobile devices becomes a concern. While open standards such as Kismet are targeted towards securing laptops, access points solutions should extend towards covering mobile devices also. Host based solutions for mobile handsets and PDA's with 802.1x interface.

Security within mobile devices fall under three categories:

- Protecting against ad hoc networks
- Connecting to rogue access points
- Mutual authentication schemes such as WPA2 as described above

Wireless IPS solutions now offer wireless security for mobile devices.

Mobile patient monitoring devices are becoming an integral part of healthcare industry and these devices will eventually become the method of choice for accessing and implementing health checks for patients located in remote areas. For these types of patient monitoring systems, security and reliability are critical, because they can influence the condition of patients, and could leave medical professionals in the dark about the condition of the patient if compromised.

Implementing Network Encryption

In order to implement 802.11i, one must first make sure both that the router/access point(s), as well as all client devices are indeed equipped to support the network encryption. If this is done, a server such as RADIUS, ADS, NDS, or LDAP needs to be integrated. This server can be a computer on the local network, an access point / router with integrated authentication server, or a remote server. AP's/routers with integrated authentication servers are often very expensive and specifically an option for commercial usage like hot spots. Hosted 802.1X servers via the Internet require a monthly fee; running a private server is free yet has the disadvantage that one must set it up and that the server needs to be on continuously.

To set up a server, server and client software must be installed. Server software required is an enterprise authentication server such as RADIUS, ADS, NDS, or LDAP. The required software can be picked from various suppliers as Microsoft, Cisco, Funk Software, Meetinghouse Data, and from some open-source projects. Software includes:

- Aradial RADIUS Server
- Cisco Secure Access Control Software
- freeRADIUS (open-source)
- Funk Software Steel Belted RADIUS (Odyssey)
- Microsoft Internet Authentication Service
- Meetinghouse Data EAGIS
- SkyFriendz (free cloud solution based on freeRADIUS)

Client software comes built-in with Windows XP and may be integrated into other OS's using any of following software:

- AEGIS-client
- Cisco ACU-client

- Intel PROSet/Wireless Software
- Odyssey client
- Xsupplicant (open1X)-project

RADIUS

Remote Authentication Dial In User Service (RADIUS) is an AAA (authentication, authorization and accounting) protocol used for remote network access. RADIUS was originally proprietary but was later published under ISOC documents RFC 2138 and RFC 2139. The idea is to have an inside server act as a gatekeeper by verifying identities through a username and password that is already pre-determined by the user. A RADIUS server can also be configured to enforce user policies and restrictions as well as record accounting information such as connection time for purposes such as billing.

Open Access Points

Today, there is almost full wireless network coverage in many urban areas - the infrastructure for the wireless community network (which some consider to be the future of the internet) is already in place. One could roam around and always be connected to Internet if the nodes were open to the public, but due to security concerns, most nodes are encrypted and the users don't know how to disable encryption. Many people consider it proper etiquette to leave access points open to the public, allowing free access to Internet. Others think the default encryption provides substantial protection at small inconvenience, against dangers of open access that they fear may be substantial even on a home DSL router.

The density of access points can even be a problem - there are a limited number of channels available, and they partly overlap. Each channel can handle multiple networks, but places with many private wireless networks (for example, apartment complexes), the limited number of Wi-Fi radio channels might cause slowness and other problems.

According to the advocates of Open Access Points, it shouldn't involve any significant risks to open up wireless networks for the public:

- The wireless network is after all confined to a small geographical area. A computer connected to the Internet and having improper configurations or other security problems can be exploited by anyone from anywhere in the world, while only clients in a small geographical range can exploit an open wireless access point. Thus the exposure is low with an open wireless access point, and the risks with having an open wireless network are small. However, one should be aware that an open wireless router will give access to the local network, often including access to file shares and printers.
- The only way to keep communication truly secure is to use end-to-end encryption. For example, when accessing an internet bank, one would almost always use strong encryption from the web browser and all the way to the bank - thus it shouldn't be risky to do banking over an unencrypted wireless network. The argument is that anyone can sniff the traffic

applies to wired networks too, where system administrators and possible hackers have access to the links and can read the traffic. Also, anyone knowing the keys for an encrypted wireless network can gain access to the data being transferred over the network.

- If services like file shares, access to printers etc. are available on the local net, it is advisable to have authentication (i.e. by password) for accessing it (one should never assume that the private network is not accessible from the outside). Correctly set up, it should be safe to allow access to the local network to outsiders.
- With the most popular encryption algorithms today, a sniffer will usually be able to compute the network key in a few minutes.
- It is very common to pay a fixed monthly fee for the Internet connection, and not for the traffic - thus extra traffic will not be detrimental.
- Where Internet connections are plentiful and cheap, freeloaders will seldom be a prominent nuisance.

On the other hand, in some countries including Germany, persons providing an open access point may be made (partially) liable for any illegal activity conducted via this access point. Also, many contracts with ISPs specify that the connection may not be shared with other persons.

Wireless Intrusion Prevention System

In computing, a wireless intrusion prevention system (WIPS) is a network device that monitors the radio spectrum for the presence of unauthorized access points (intrusion detection), and can automatically take countermeasures (intrusion prevention).

Purpose

The primary purpose of a WIPS is to prevent unauthorized network access to local area networks and other information assets by wireless devices. These systems are typically implemented as an overlay to an existing Wireless LAN infrastructure, although they may be deployed standalone to enforce no-wireless policies within an organization. Some advanced wireless infrastructure has integrated WIPS capabilities.

Large organizations with many employees are particularly vulnerable to security breaches caused by rogue access points. If an employee (trusted entity) in a location brings in an easily available wireless router, the entire network can be exposed to anyone within range of the signals.

In July 2009, the PCI Security Standards Council published wireless guidelines for PCI DSS recommending the use of WIPS to automate wireless scanning for large organizations.

Intrusion Detection

A wireless intrusion detection system (WIDS) monitors the radio spectrum for the presence of unauthorized, rogue access points and the use of wireless attack tools. The system monitors the radio spectrum used by wireless LANs, and immediately alerts a systems administrator whenever a rogue access point is detected. Conventionally it is achieved by comparing the MAC address of the participating wireless devices.

Rogue devices can spoof MAC address of an authorized network device as their own. New research uses fingerprinting approach to weed out devices with spoofed MAC addresses. The idea is to compare the unique signatures exhibited by the signals emitted by each wireless device against the known signatures of pre-authorized, known wireless devices.

Intrusion Prevention

In addition to intrusion detection, a WIPS also includes features that prevent against the threat *automatically*. For automatic prevention, it is required that the WIPS is able to accurately detect and automatically classify a threat.

The following types of threats can be prevented by a good WIPS:

- Rogue access points – WIPS should understand the difference between rogue APs and external (neighbor's) APs
- Mis-configured AP
- Client mis-association
- Unauthorized association
- Man-in-the-middle attack
- *Ad hoc* networks
- MAC spoofing
- Honeypot / evil twin attack
- Denial-of-service attack

Implementation

WIPS configurations consist of three components:

- Sensors — These devices contain antennas and radios that scan the wireless spectrum for packets and are installed throughout areas to be protected.
- Server — The WIPS server centrally analyzes packets captured by sensors.
- Console — The console provides the primary user interface into the system for administration and reporting.

A simple intrusion detection system can be a single computer, connected to a wireless signal processing device, and antennas placed throughout the facility. For huge organizations, a Multi Network Controller provides central control of multiple WIPS servers, while for SOHO or SMB customers, all the functionality of WIPS is available in single box.

In a WIPS implementation, users first define the operating wireless policies in the WIPS. The

WIPS sensors then analyze the traffic in the air and send this information to WIPS server. The WIPS server correlates the information, validates it against the defined policies, and classifies if it is a threat. The administrator of the WIPS is then notified of the threat, or, if a policy has been set accordingly, the WIPS takes automatic protection measures.

WIPS is configured as either a network implementation or a hosted implementation.

Network Implementation

In a network WIPS implementation, server, sensors and the console are all placed inside a private network and are not accessible from the Internet.

Sensors communicate with the server over a private network using a private port. Since the server resides on the private network, users can access the console only from within the private network.

A network implementation is suitable for organizations where all locations are within the private network.

Hosted Implementation

In a hosted WIPS implementation, sensors are installed inside a private network. However, the server is hosted in secure data center and is accessible on the Internet. Users can access the WIPS console from anywhere on the Internet. A hosted WIPS implementation is as secure as a network implementation because the data flow is encrypted between sensors and server, as well as between server and console. A hosted WIPS implementation requires very little configuration because the sensors are programmed to automatically look for the server on the Internet over a secure TLS connection.

For a large organization with locations that are not a part of a private network, a hosted WIPS implementation simplifies deployment significantly because sensors connect to the Server over the Internet without requiring any special configuration. Additionally, the Console can be accessed securely from anywhere on the Internet.

Hosted WIPS implementations are available in an on-demand, subscription-based software as a service model. Hosted implementations may be appropriate for organizations looking to fulfill the minimum scanning requirements of PCI DSS.

Intrusion Detection System

An intrusion detection system (IDS) is a device or software application that monitors a network or systems for malicious activity or policy violations. Any detected activity or violation is typically reported either to an administrator or collected centrally using a security information and event management (SIEM) system. A SIEM system combines outputs from multiple sources, and uses alarm filtering techniques to distinguish malicious activity from false alarms.

There is a wide spectrum of IDS, varying from antivirus software to hierarchical systems that monitor the traffic of an entire backbone network. The most common classifications are network intrusion detection systems (NIDS) and host-based intrusion detection systems (HIDS). A system that monitors important operating system files is an example of a HIDS, while a system that analyzes incoming network traffic is an example of a NIDS. It is also possible to classify IDS by detection approach: the most well-known variants are signature-based detection (recognizing bad patterns, such as malware) and anomaly-based detection (detecting deviations from a model of "good" traffic, which often relies on machine learning). Some IDS have the ability to respond to detected intrusions. Systems with response capabilities are typically referred to as an intrusion prevention system.

Comparison with Firewalls

Though they both relate to network security, an intrusion detection system (IDS) differs from a firewall in that a firewall looks outwardly for intrusions in order to stop them from happening. Firewalls limit access between networks to prevent intrusion and do not signal an attack from inside the network. An IDS evaluates a suspected intrusion once it has taken place and signals an alarm. An IDS also watches for attacks that originate from within a system. This is traditionally achieved by examining network communications, identifying heuristics and patterns (often known as signatures) of common computer attacks, and taking action to alert operators. A system that terminates connections is called an intrusion prevention system, and is another form of an application layer firewall.

Classifications

IDS can be classified by where detection takes place (network or host) and the detection method that is employed.

Analyzed Activity

Network Intrusion Detection Systems

Network intrusion detection systems (NIDS) are placed at a strategic point or points within the network to monitor traffic to and from all devices on the network. It performs an analysis of passing traffic on the entire subnet, and matches the traffic that is passed on the subnets to the library of known attacks. Once an attack is identified, or abnormal behavior is sensed, the alert can be sent to the administrator. An example of an NIDS would be installing it on the subnet where firewalls are located in order to see if someone is trying to break into the firewall. Ideally one would scan all inbound and outbound traffic, however doing so might create a bottleneck that would impair the overall speed of the network. OPNET and NetSim are commonly used tools for simulation network intrusion detection systems. NID Systems are also capable of comparing signatures for similar packets to link and drop harmful detected packets which have a signature matching the records in the NIDS. When we classify the designing of the NIDS according to the system interactivity property, there are two types: on-line and off-line NIDS. On-line NIDS deals with the network in real time. It analyses the Ethernet packets and applies some rules, to decide if it is an attack or not. Off-line NIDS deals with stored data and passes it through some processes to decide if it is an attack or not.

Host Intrusion Detection Systems

Host intrusion detection systems (HIDS) run on individual hosts or devices on the network. A

HIDS monitors the inbound and outbound packets from the device only and will alert the user or administrator if suspicious activity is detected. It takes a snapshot of existing system files and matches it to the previous snapshot. If the critical system files were modified or deleted, an alert is sent to the administrator to investigate. An example of HIDS usage can be seen on mission critical machines, which are not expected to change their configurations.

Intrusion detection systems can also be system-specific using custom tools and honeypots.

Detection Method

Signature-based

Signature-based IDS refers to the detection of attacks by looking for specific patterns, such as byte sequences in network traffic, or known malicious instruction sequences used by malware. This terminology originates from anti-virus software, which refers to these detected patterns as signatures. Although signature-based IDS can easily detect known attacks, it is impossible to detect new attacks, for which no pattern is available.

Anomaly-based

Anomaly-based intrusion detection systems were primarily introduced to detect unknown attacks, in part due to the rapid development of malware. The basic approach is to use machine learning to create a model of trustworthy activity, and then compare new behavior against this model. Although this approach enables the detection of previously unknown attacks, it suffers from false positives: previously unknown legitimate activity may also be classified as malicious.

Intrusion Prevention

Some systems may attempt to stop an intrusion attempt but this is neither required nor expected of a monitoring system. Intrusion detection and prevention systems (IDPS) are primarily focused on identifying possible incidents, logging information about them, and reporting attempts. In addition, organizations use IDPSes for other purposes, such as identifying problems with security policies, documenting existing threats and deterring individuals from violating security policies. IDPSes have become a necessary addition to the security infrastructure of nearly every organization.

IDPSes typically record information related to observed events, notify security administrators of important observed events and produce reports. Many IDPSes can also respond to a detected threat by attempting to prevent it from succeeding. They use several response techniques, which involve the IDPS stopping the attack itself, changing the security environment (e.g. reconfiguring a firewall) or changing the attack's content.

Intrusion prevention systems (IPS), also known as intrusion detection and prevention systems (IDPS), are network security appliances that monitor network or system activities for malicious activity. The main functions of intrusion prevention systems are to identify malicious activity, log information about this activity, report it and attempt to block or stop it.

Intrusion prevention systems are considered extensions of intrusion detection systems because they both monitor network traffic and/or system activities for malicious activity. The main differences are, unlike intrusion detection systems, intrusion prevention systems are placed in-line and are able to actively prevent or block intrusions that are detected. IPS can take such actions as sending an alarm, dropping detected malicious packets, resetting a connection or blocking traffic from the offending IP address. An IPS also can correct cyclic redundancy check (CRC) errors, defragment packet streams, mitigate TCP sequencing issues, and clean up unwanted transport and network layer options.

Classification

Intrusion prevention systems can be classified into four different types:

- Network-based intrusion prevention system (NIPS): monitors the entire network for suspicious traffic by analyzing protocol activity.
- Wireless intrusion prevention systems (WIPS): monitor a wireless network for suspicious traffic by analyzing wireless networking protocols.
- Network behavior analysis (NBA): examines network traffic to identify threats that generate unusual traffic flows, such as distributed denial of service (DDoS) attacks, certain forms of malware and policy violations.
- Host-based intrusion prevention system (HIPS): an installed software package which monitors a single host for suspicious activity by analyzing events occurring within that host.

Detection Methods

The majority of intrusion prevention systems utilize one of three detection methods: signature-based, statistical anomaly-based, and stateful protocol analysis.

- Signature-Based Detection: Signature based IDS monitors packets in the Network and compares with pre-configured and pre-determined attack patterns known as signatures.
- Statistical anomaly-based detection: An IDS which is anomaly based will monitor network traffic and compare it against an established baseline. The baseline will identify what is "normal" for that network – what sort of bandwidth is generally used, what protocols are used that it may raise a False Positive alarm for a legitimate use of bandwidth if the baselines are not intelligently configured.
- Stateful Protocol Analysis Detection: This method identifies deviations of protocol states by comparing observed events with "predetermined profiles of generally accepted definitions of benign activity."

Limitations

- Noise can severely limit an intrusion detection system's effectiveness. Bad packets generated from software bugs, corrupt DNS data, and local packets that escaped can create a significantly high false-alarm rate.

- It is not uncommon for the number of real attacks to be far below the number of false-alarms. Number of real attacks is often so far below the number of false-alarms that the real attacks are often missed and ignored.
- Many attacks are geared for specific versions of software that are usually outdated. A constantly changing library of signatures is needed to mitigate threats. Outdated signature databases can leave the IDS vulnerable to newer strategies.
- For signature-based IDSes there will be lag between a new threat discovery and its signature being applied to the IDS. During this lag time the IDS will be unable to identify the threat.
- It cannot compensate for a weak identification and authentication mechanisms or for weaknesses in network protocols. When an attacker gains access due to weak authentication mechanism then IDS cannot prevent the adversary from any malpractice.
- Encrypted packets are not processed by the intrusion detection software. Therefore, the encrypted packet can allow an intrusion to the network that is undiscovered until more significant network intrusions have occurred.
- Intrusion detection software provides information based on the network address that is associated with the IP packet that is sent into the network. This is beneficial if the network address contained in the IP packet is accurate. However, the address that is contained in the IP packet could be faked or scrambled.
- Due to the nature of NIDS systems, and the need for them to analyse protocols as they are captured, NIDS systems can be susceptible to same protocol based attacks that network hosts may be vulnerable. Invalid data and TCP/IP stack attacks may cause an NIDS to crash.

Evasion Techniques

There are a number of techniques which attackers are using, the following are considered 'simple' measures which can be taken to evade IDS:

- Fragmentation: by sending fragmented packets, the attacker will be under the radar and can easily bypass the detection system's ability to detect the attack signature.
- Avoiding defaults: The TCP port utilised by a protocol does not always provide an indication to the protocol which is being transported. For example, an IDS may expect to detect a trojan on port 12345. If an attacker had reconfigured it to use a different port the IDS may not be able to detect the presence of the trojan.
- Coordinated, low-bandwidth attacks: coordinating a scan among numerous attackers (or agents) and allocating different ports or hosts to different attackers makes it difficult for the IDS to correlate the captured packets and deduce that a network scan is in progress.
- Address spoofing/proxying: attackers can increase the difficulty of the ability of Security Administrators to determine the source of the attack by using poorly secured or incorrectly configured proxy servers to bounce an attack. If the source is spoofed and bounced by a server then it makes it very difficult for IDS to detect the origin of the attack.

- Pattern change evasion: IDSs generally rely on 'pattern matching' to detect an attack. By changing the data used in the attack slightly, it may be possible to evade detection. For example, an IMAP server may be vulnerable to a buffer overflow, and an IDS is able to detect the attack signature of 10 common attack tools. By modifying the payload sent by the tool, so that it does not resemble the data that the IDS expects, it may be possible to evade detection.

Development

One preliminary IDS concept consisted of a set of tools intended to help administrators review audit trails. User access logs, file access logs, and system event logs are examples of audit trails.

Fred Cohen noted in 1987 that it is impossible to detect an intrusion in every case, and that the resources needed to detect intrusions grow with the amount of usage.

Dorothy E. Denning, assisted by Peter G. Neumann, published a model of an IDS in 1986 that formed the basis for many systems today. Her model used statistics for anomaly detection, and resulted in an early IDS at SRI International named the Intrusion Detection Expert System (IDES), which ran on Sun workstations and could consider both user and network level data. IDES had a dual approach with a rule-based Expert System to detect known types of intrusions plus a statistical anomaly detection component based on profiles of users, host systems, and target systems. Lunt proposed adding an Artificial neural network as a third component. She said all three components could then report to a resolver. SRI followed IDES in 1993 with the Next-generation Intrusion Detection Expert System (NIDES).

The Multics intrusion detection and alerting system (MIDAS), an expert system using P-BEST and Lisp, was developed in 1988 based on the work of Denning and Neumann. Haystack was also developed in that year using statistics to reduce audit trails.

Wisdom & Sense (W&S) was a statistics-based anomaly detector developed in 1989 at the Los Alamos National Laboratory. W&S created rules based on statistical analysis, and then used those rules for anomaly detection.

In 1990, the Time-based Inductive Machine (TIM) did anomaly detection using inductive learning of sequential user patterns in Common Lisp on a VAX 3500 computer. The Network Security Monitor (NSM) performed masking on access matrices for anomaly detection on a Sun-3/50 workstation. The Information Security Officer's Assistant (ISOA) was a 1990 prototype that considered a variety of strategies including statistics, a profile checker, and an expert system. ComputerWatch at AT&T Bell Labs used statistics and rules for audit data reduction and intrusion detection.

Then, in 1991, researchers at the University of California, Davis created a prototype Distributed Intrusion Detection System (DIDS), which was also an expert system. The Network Anomaly Detection and Intrusion Reporter (NADIR), also in 1991, was a prototype IDS developed at the Los Alamos National Laboratory's Integrated Computing Network (ICN), and was heavily influenced by the work of Denning and Lunt. NADIR used a statistics-based anomaly detector and an expert system.

The Lawrence Berkeley National Laboratory announced Bro in 1998, which used its own rule language for packet analysis from libpcap data. Network Flight Recorder (NFR) in 1999 also used libpcap. APE was developed as a packet sniffer, also using libpcap, in November, 1998, and was renamed Snort one month later. APE has since become the world's largest used IDS/IPS system with over 300,000 active users.

The Audit Data Analysis and Mining (ADAM) IDS in 2001 used tcpdump to build profiles of rules for classifications.

In 2003, Yongguang Zhang and Wenke Lee argue for the importance of IDS in networks with mobile nodes.

RADIUS

Remote Authentication Dial-In User Service (RADIUS) is a networking protocol that provides centralized Authentication, Authorization, and Accounting (AAA or Triple A) management for users who connect and use a network service. RADIUS was developed by Livingston Enterprises, Inc. in 1991 as an access server authentication and accounting protocol and later brought into the Internet Engineering Task Force (IETF) standards.

Because of the broad support and the ubiquitous nature of the RADIUS protocol, it is often used by ISPs and enterprises to manage access to the Internet or internal networks, wireless networks, and integrated e-mail services. These networks may incorporate modems, DSL, access points, VPNs, network ports, web servers, etc.

RADIUS is a client/server protocol that runs in the application layer, and can use either TCP or UDP as transport. Network access servers, the gateways that control access to a network, usually contain a RADIUS client component that communicates with the RADIUS server. RADIUS is often the back-end of choice for 802.1X authentication as well.

The RADIUS server is usually a background process running on a UNIX or Microsoft Windows server.

Protocol Components

RADIUS is an AAA protocol which manages network access in the following two-step process, also known as a *AAA transaction*. AAA stands for authentication, authorization and accounting. Authentication and authorization characteristics in RADIUS are described in RFC 2865 while accounting is described by RFC 2866.

Authentication and Authorization

The user or machine sends a request to a Network Access Server (NAS) to gain access to a particular network resource using access credentials. The credentials are passed to the NAS device via the link-layer protocol - for example, Point-to-Point Protocol (PPP) in the case of many dialup or DSL providers or posted in an HTTPS secure web form.

In turn, the NAS sends a RADIUS *Access Request* message to the RADIUS server, requesting authorization to grant access via the RADIUS protocol.

This request includes access credentials, typically in the form of username and password or security certificate provided by the user. Additionally, the request may contain other information which the NAS knows about the user, such as its network address or phone number, and information regarding the user's physical point of attachment to the NAS.

The RADIUS server checks that the information is correct using authentication schemes such as PAP, CHAP or EAP. The user's proof of identification is verified, along with, optionally, other information related to the request, such as the user's network address or phone number, account status, and specific network service access privileges. Historically, RADIUS servers checked the user's information against a locally stored flat file database. Modern RADIUS servers can do this, or can refer to external sources — commonly SQL, Kerberos, LDAP, or Active Directory servers — to verify the user's credentials.

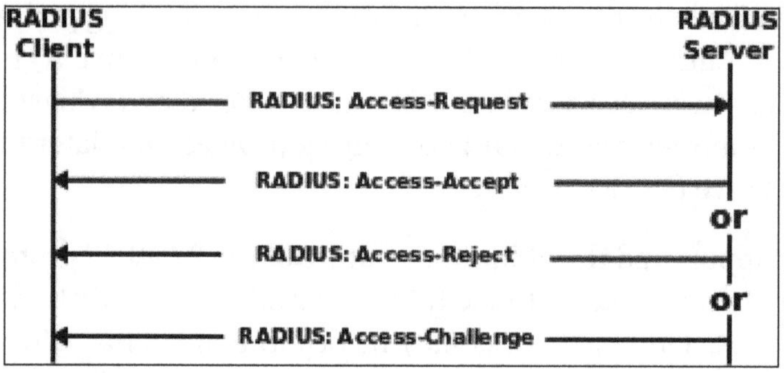

RADIUS Authentication and Authorization Flow:

The RADIUS server then returns one of three responses to the NAS : 1) Access Reject, 2) Access Challenge, or 3) Access Accept.

- *Access Reject* - The user is unconditionally denied access to all requested network resources. Reasons may include failure to provide proof of identification or an unknown or inactive user account.
- *Access Challenge* - Requests additional information from the user such as a secondary password, PIN, token, or card. Access Challenge is also used in more complex authentication dialogs where a secure tunnel is established between the user machine and the Radius Server in a way that the access credentials are hidden from the NAS.
- *Access Accept* - The user is granted access. Once the user is authenticated, the RADIUS server will often check that the user is authorized to use the network service requested. A given user may be allowed to use a company's wireless network, but not its VPN service, for example. Again, this information may be stored locally on the RADIUS server, or may be looked up in an external source such as LDAP or Active Directory.

Each of these three RADIUS responses may include a Reply-Message attribute which may give a reason for the rejection, the prompt for the challenge, or a welcome message for the accept. The text in the attribute can be passed on to the user in a return web page.

Authorization attributes are conveyed to the NAS stipulating terms of access to be granted. For example, the following authorization attributes may be included in an Access-Accept:

- The specific IP address to be assigned to the user.

- The address pool from which the user's IP should be chosen.
- The maximum length of time that the user may remain connected.
- An access list, priority queue or other restrictions on a user's access.
- L2TP parameters.
- VLAN parameters.
- Quality of Service (QoS) parameters.

When a client is configured to use RADIUS, any user of the client presents authentication information to the client. This might be with a customizable login prompt, where the user is expected to enter their username and password. Alternatively, the user might use a link framing protocol such as the Point-to-Point Protocol (PPP), which has authentication packets which carry this information.

Once the client has obtained such information, it may choose to authenticate using RADIUS. To do so, the client creates an "Access- Request" containing such Attributes as the user's name, the user's password, the ID of the client and the Port ID which the user is accessing. When a password is present, it is hidden using a method based on the RSA Message Digest Algorithm MD5.

Accounting

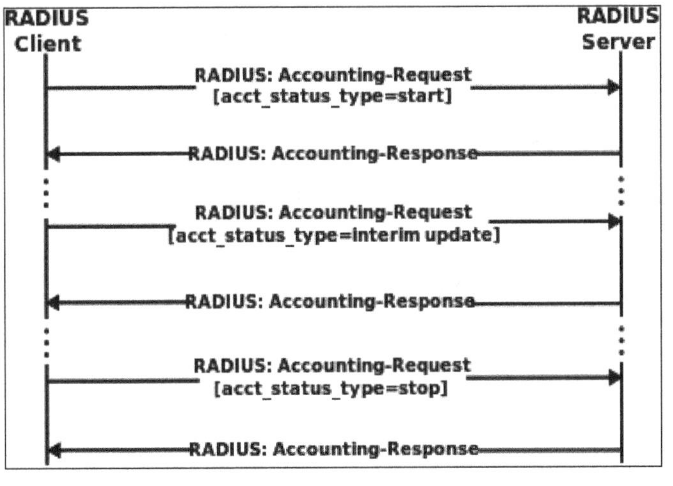

RADIUS Accounting Flow.

Accounting is described in RFC 2866.

When network access is granted to the user by the NAS, an *Accounting Start* (a RADIUS Accounting Request packet containing an Acct-Status-Type attribute with the value "start") is sent by the NAS to the RADIUS server to signal the start of the user's network access. "Start" records typically contain the user's identification, network address, point of attachment and a unique session identifier.

Periodically, *Interim Update* records (a RADIUS Accounting Request packet containing an Acct-Status-Type attribute with the value "interim-update") may be sent by the NAS to the RADIUS server, to update it on the status of an active session. "Interim" records typically convey the current session duration and information on current data usage.

Finally, when the user's network access is closed, the NAS issues a final *Accounting Stop* record (a RADIUS Accounting Request packet containing an Acct-Status-Type attribute with the value "stop") to the RADIUS server, providing information on the final usage in terms of time, packets transferred, data transferred, reason for disconnect and other information related to the user's network access.

Typically, the client sends Accounting-Request packets until it receives an Accounting-Response acknowledgement, using some retry interval.

The primary purpose of this data is that the user can be billed accordingly; the data is also commonly used for statistical purposes and for general network monitoring.

Roaming

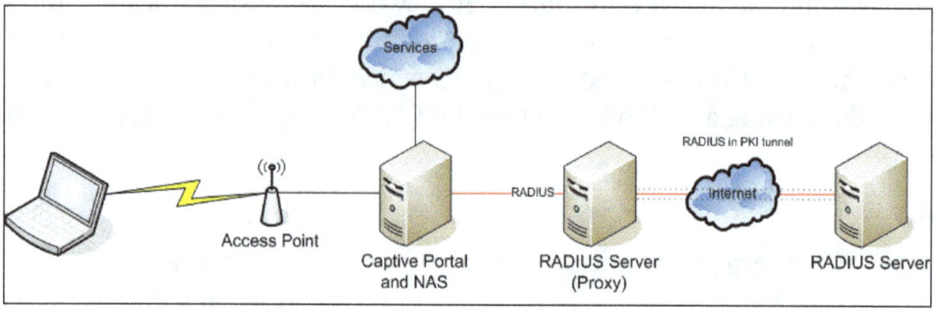

Roaming using a proxy RADIUS AAA server.

RADIUS is commonly used to facilitate roaming between ISPs, for example:

- By companies which provide a single global set of credentials that are usable on many public networks.
- By independent, but collaborating, institutions issuing their own credentials to their own users, that allow a visitor from one to another to be authenticated by their home institution, such as in eduroam.

RADIUS facilitates this by the use of *realms*, which identify where the RADIUS server should forward the AAA requests for processing.

Realms

A realm is commonly appended to a user's user name and delimited with an '@' sign, resembling an email address domain name. This is known as *postfix* notation for the realm. Another common usage is *prefix* notation, which involves prepending the realm to the username and using '\' as a delimiter. Modern RADIUS servers allow any character to be used as a realm delimiter, although in practice '@' and '\' are usually used.

Realms can also be compounded using both prefix and postfix notation, to allow for complicated roaming scenarios; for example, somedomain.com\username@anotherdomain.com could be a valid username with two realms.

Although realms often resemble domains, it is important to note that realms are in fact arbitrary text and need not contain real domain names. Realm formats are standardized in RFC 4282, which defines a Network Access Identifier (NAI) in the form of 'user@realm'. In that specification, the 'realm' portion is required to be a domain name. However, this practice is not always followed. RFC 7542 replaced RFC 4282 in May 2015.

Proxy Operations

When a RADIUS server receives an AAA request for a user name containing a realm, the server will reference a table of configured realms. If the realm is known, the server will then *proxy* the request to the configured home server for that domain. The behavior of the proxying server regarding the removal of the realm from the request ("stripping") is configuration-dependent on most servers. In addition, the proxying server can be configured to add, remove or rewrite AAA requests when they are proxied over time again.

Proxy Chaining is possible in RADIUS and authentication/authorization and accounting packets are usually routed between a NAS Device and a Home server through a series of proxies. Some of advantages of using Proxy chains include scalability improvements, policy implementations and capability adjustments. But in roaming scenarios, the NAS, Proxies and Home Server could be typically managed by different administrative entities. Hence, the trust factor among the proxies gains more significance under such Inter-domain applications. Further, the absence of end to end security in RADIUS adds to the criticality of trust among the Proxies involved. Proxy Chains are explained in RFC 2607.

Security

Roaming with RADIUS exposes the users to various security and privacy concerns. More generally, some roaming partners establish a secure tunnel between the RADIUS servers to ensure that users' credentials cannot be intercepted while being proxied across the internet. This is a concern as the MD5 hash built into RADIUS is considered insecure.

Packet Structure

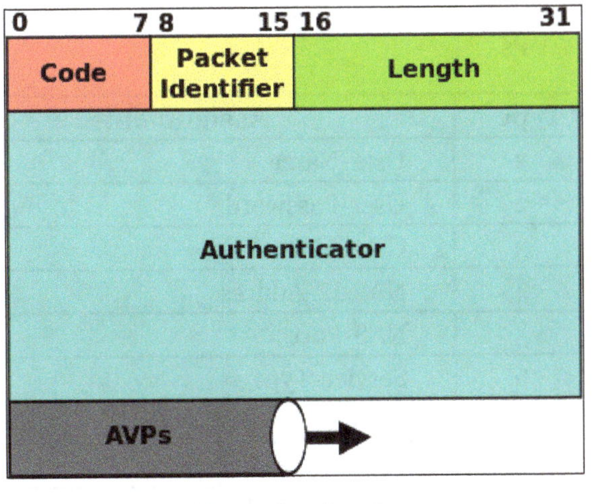

RADIUS packet data format.

The RADIUS packet data format is shown to the right. The fields are transmitted from left to right, starting with the code, the identifier, the length, the authenticator and the attributes.

RADIUS Codes (decimal) are assigned as follows:

Code	Assignment
1	Access-Request
2	Access-Accept
3	Access-Reject
4	Accounting-Request
5	Accounting-Response
11	Access-Challenge
12	Status-Server (experimental)
13	Status-Client (experimental)
255	Reserved

The Identifier field aids in matching requests and replies.

The Length field indicates the length of the entire RADIUS packet including the Code, Identifier, Length, Authenticator and optional Attribute fields.

The Authenticator is used to authenticate the reply from the RADIUS server, and is used in encrypting passwords; its length is 16 bytes.

Attribute Value Pairs

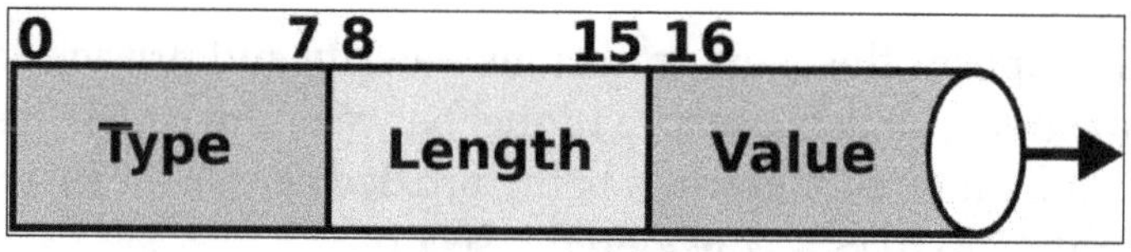

RADIUS AVP layout.

The RADIUS Attribute Value Pairs (AVP) carry data in both the request and the response for the authentication, authorization, and accounting transactions. The length of the radius packet is used to determine the end of the AVPs.

AVP Type	Assignment
1	User-Name
2	User-Password
3	CHAP-Password
4	NAS-IP-Address
5	NAS-Port
6	Service-Type
7	Framed-Protocol
8	Framed-IP-Address
9	Framed-IP-Netmask

10	Framed-Routing
11	Filter-Id
12	Framed-MTU
13	Framed-Compression
14	Login-IP-Host
15	Login-Service
16	Login-TCP-Port
18	Reply-Message
19	Callback-Number
20	Callback-Id
22	Framed-Route
23	Framed-IPX-Network
24	State
25	Class
26	Vendor-Specific
27	Session-Timeout
28	Idle-Timeout
29	Termination-Action
30	Called-Station-Id
31	Calling-Station-Id
32	NAS-Identifier
33	Proxy-State
34	Login-LAT-Service
35	Login-LAT-Node
36	Login-LAT-Group
37	Framed-AppleTalk-Link
38	Framed-AppleTalk-Network
39	Framed-AppleTalk-Zone
40	Acct-Status-Type
41	Acct-Delay-Time
42	Acct-Input-Octets
43	Acct-Output-Octets
44	Acct-Session-Id
45	Acct-Authentic
46	Acct-Session-Time
47	Acct-Input-Packets
48	Acct-Output-Packets
49	Acct-Terminate-Cause
50	Acct-Multi-Session-Id
51	Acct-Link-Count

52	Acct-Input-Gigawords
53	Acct-Output-Gigawords
55	Event-Timestamp
56	Egress-VLANID
57	Ingress-Filters
58	Egress-VLAN-Name
59	User-Priority-Table
60	CHAP-Challenge
61	NAS-Port-Type
62	Port-Limit
63	Login-LAT-Port
64	Tunnel-Type
65	Tunnel-Medium-Type
66	Tunnel-Client-Endpoint
67	Tunnel-Server-Endpoint
68	Acct-Tunnel-Connection
69	Tunnel-Password
70	ARAP-Password
71	ARAP-Features
72	ARAP-Zone-Access
73	ARAP-Security
74	ARAP-Security-Data
75	Password-Retry
76	Prompt
77	Connect-Info
78	Configuration-Token
79	EAP-Message
80	Message-Authenticator
81	Tunnel-Private-Group-ID
82	Tunnel-Assignment-ID
83	Tunnel-Preference
84	ARAP-Challenge-Response
85	Acct-Interim-Interval
86	Acct-Tunnel-Packets-Lost
87	NAS-Port-Id
88	Framed-Pool
89	CUI
90	Tunnel-Client-Auth-ID
91	Tunnel-Server-Auth-ID
92	NAS-Filter-Rule

94	Originating-Line-Info
95	NAS-IPv6-Address
96	Framed-Interface-Id
97	Framed-IPv6-Prefix
98	Login-IPv6-Host
99	Framed-IPv6-Route
100	Framed-IPv6-Pool
101	Error-Cause Attribute
102	EAP-Key-Name
103	Digest-Response
104	Digest-Realm
105	Digest-Nonce
106	Digest-Response-Auth
107	Digest-Nextnonce
108	Digest-Method
109	Digest-URI
110	Digest-Qop
111	Digest-Algorithm
112	Digest-Entity-Body-Hash
113	Digest-CNonce
114	Digest-Nonce-Count
115	Digest-Username
116	Digest-Opaque
117	Digest-Auth-Param
118	Digest-AKA-Auts
119	Digest-Domain
120	Digest-Stale
121	Digest-HA1
122	SIP-AOR
123	Delegated-IPv6-Prefix
124	MIP6-Feature-Vector
125	MIP6-Home-Link-Prefix
126	Operator-Name
127	Location-Information
128	Location-Data
129	Basic-Location-Policy-Rules
130	Extended-Location-Policy-Rules
131	Location-Capable
132	Requested-Location-Info
133	Framed-Management-Protocol

134	Management-Transport-Protection
135	Management-Policy-Id
136	Management-Privilege-Level
137	PKM-SS-Cert
138	PKM-CA-Cert
139	PKM-Config-Settings
140	PKM-Cryptosuite-List
141	PKM-SAID
142	PKM-SA-Descriptor
143	PKM-Auth-Key
144	DS-Lite-Tunnel-Name
145	Mobile-Node-Identifier
146	Service-Selection
147	PMIP6-Home-LMA-IPv6-Address
148	PMIP6-Visited-LMA-IPv6-Address
149	PMIP6-Home-LMA-IPv4-Address
150	PMIP6-Visited-LMA-IPv4-Address
151	PMIP6-Home-HN-Prefix
152	PMIP6-Visited-HN-Prefix
153	PMIP6-Home-Interface-ID
154	PMIP6-Visited-Interface-ID
155	PMIP6-Home-IPv4-HoA
156	PMIP6-Visited-IPv4-HoA
157	PMIP6-Home-DHCP4-Server-Address
158	PMIP6-Visited-DHCP4-Server-Address
159	PMIP6-Home-DHCP6-Server-Address
160	PMIP6-Visited-DHCP6-Server-Address
161	PMIP6-Home-IPv4-Gateway
162	PMIP6-Visited-IPv4-Gateway
163	EAP-Lower-Layer
164	GSS-Acceptor-Service-Name
165	GSS-Acceptor-Host-Name
166	GSS-Acceptor-Service-Specifics
167	GSS-Acceptor-Realm-Name
168	Framed-IPv6-Address
169	DNS-Server-IPv6-Address
170	Route-IPv6-Information
171	Delegated-IPv6-Prefix-Pool
172	Stateful-IPv6-Address-Pool
173	IPv6-6rd-Configuration

174	Allowed-Called-Station-Id
175	EAP-Peer-Id
176	EAP-Server-Id
177	Mobility-Domain-Id
178	Preauth-Timeout
179	Network-Id-Name
180	EAPoL-Announcement
181	WLAN-HESSID
182	WLAN-Venue-Info
183	WLAN-Venue-Language
184	WLAN-Venue-Name
185	WLAN-Reason-Code
186	WLAN-Pairwise-Cipher
187	WLAN-Group-Cipher
188	WLAN-AKM-Suite
189	WLAN-Group-Mgmt-Cipher
190	WLAN-RF-Band

Vendor-specific Attributes

RADIUS is extensible; many vendors of RADIUS hardware and software implement their own variants using Vendor-Specific Attributes (VSAs). Microsoft has published some of their VSAs. VSA definitions from many other companies remain proprietary and/or ad-hoc, nonetheless many VSA dictionaries can be found by downloading the source code of open source RADIUS implementations, for example FreeRADIUS or openRADIUS.

Security

The RADIUS protocol transmits obfuscated passwords using a shared secret and the MD5 hashing algorithm. As this particular implementation provides only weak protection of the user's credentials, additional protection, such as IPsec tunnels or physically secured data-center networks, should be used to further protect the RADIUS traffic between the NAS device and the RADIUS server. Additionally, the user's security credentials are the only part protected by RADIUS itself, yet other user-specific attributes such as tunnel-group IDs or vlan memberships passed over RADIUS may be considered sensitive (helpful to an attacker) or private (sufficient to identify the individual client) information as well. The RadSec protocol claims to solve aforementioned security issues.

History

RADIUS was originally specified in an RFI by Merit Network in 1991 to control dial-in access to NSFnet. Livingston Enterprises responded to the RFI with a description of a RADIUS server. Merit Network awarded the contract to Livingston Enterprises that delivered their PortMaster series of Network Access Servers and the initial RADIUS server to Merit. RADIUS was later (1997) published as RFC 2058 and RFC 2059 (current versions are RFC 2865 and RFC 2866).

Now, several commercial and open-source RADIUS servers exist. Features can vary, but most can look up the users in text files, LDAP servers, various databases, etc. Accounting records can be written to text files, various databases, forwarded to external servers, etc. SNMP is often used for remote monitoring and keep-alive checking of a RADIUS server. RADIUS proxy servers are used for centralized administration and can rewrite RADIUS packets on the fly (for security reasons, or to convert between vendor dialects).

The Diameter protocol was intended as the replacement for RADIUS. While both are Authentication, Authorization, and Accounting (AAA) protocols, the use-cases for the two protocols have since diverged. Diameter is largely used in the 3G space. RADIUS is used elsewhere. One of the largest barriers to having Diameter replace RADIUS is that switches and Access Points typically implement RADIUS, but not Diameter.

Diameter uses SCTP or TCP while RADIUS typically uses UDP as the transport layer. As of 2012, RADIUS can also use TCP as the transport layer with TLS for security.

Wi-Fi Protected Access

An example of a Wi-Fi Protected Access label found on a consumer device.

Wi-Fi Protected Access (WPA) and Wi-Fi Protected Access II (WPA2) are two security protocols and security certification programs developed by the Wi-Fi Alliance to secure wireless computer networks. The Alliance defined these in response to serious weaknesses researchers had found in the previous system, Wired Equivalent Privacy (WEP).

WPA (sometimes referred to as the *draft IEEE 802.11i* standard) became available in 2003. The Wi-Fi Alliance intended it as an intermediate measure in anticipation of the availability of the more secure and complex WPA2. WPA2 became available in 2004 and is a common shorthand for the full IEEE 802.11i (or IEEE 802.11i-2004) standard.

A flaw in a feature added to Wi-Fi, called Wi-Fi Protected Setup, allows WPA and WPA2 security to be bypassed and effectively broken in many situations. The WPA and WPA2 security protocols implemented without using the Wi-Fi Protected Setup feature are unaffected by the security vulnerability.

WPA

The Wi-Fi Alliance intended WPA as an intermediate measure to take the place of WEP pending the availability of the full IEEE 802.11i standard. WPA could be implemented through firmware upgrades on wireless network interface cards designed for WEP that began shipping as far back as 1999. However, since the changes required in the wireless access points (APs) were more extensive than those needed on the network cards, most pre-2003 APs could not be upgraded to support WPA.

The WPA protocol implements much of the IEEE 802.11i standard. Specifically, the Temporal Key Integrity Protocol (TKIP) was adopted for WPA. WEP used a 64-bit or 128-bit encryption key that must be manually entered on wireless access points and devices and does not change. TKIP employs a per-packet key, meaning that it dynamically generates a new 128-bit key for each packet and thus prevents the types of attacks that compromised WEP.

WPA also includes a message integrity check, which is designed to prevent an attacker from altering and resending data packets. This replaces the cyclic redundancy check (CRC) that was used by the WEP standard. CRC's main flaw was that it did not provide a sufficiently strong data integrity guarantee for the packets it handled. Well tested message authentication codes existed to solve these problems, but they required too much computation to be used on old network cards. WPA uses a message integrity check algorithm called *TKIP* to verify the integrity of the packets. TKIP is much stronger than a CRC, but not as strong as the algorithm used in WPA2. Researchers have since discovered a flaw in WPA that relied on older weaknesses in WEP and the limitations of Michael to retrieve the keystream from short packets to use for re-injection and spoofing.

WPA2

WPA2 replaced WPA. WPA2, which requires testing and certification by the Wi-Fi Alliance, implements the mandatory elements of IEEE 802.11i. In particular, it includes mandatory support for CCMP, an AES-based encryption mode with strong security. Certification began in September, 2004; from March 13, 2006, WPA2 certification is mandatory for all new devices to bear the Wi-Fi trademark.

Hardware Support

WPA has been designed specifically to work with wireless hardware produced prior to the introduction of WPA protocol, which provides inadequate security through WEP. Some of these devices support WPA only after applying firmware upgrades, which are not available for some legacy devices.

Wi-Fi devices certified since 2006 support both the WPA and WPA2 security protocols. WPA2 may not work with some older network cards.

WPA Terminology

Different WPA versions and protection mechanisms can be distinguished based on the target end-user (according to the method of authentication key distribution), and the encryption protocol used.

Target Users (Authentication Key Distribution)

WPA-Personal:

Also referred to as *WPA-PSK* (pre-shared key) mode, this is designed for home and small office networks and doesn't require an authentication server. Each wireless network device encrypts the network traffic using a 256 bit key. This key may be entered either as a string of 64 hexadecimal digits, or as a passphrase of 8 to 63 printable ASCII characters. If ASCII characters are used, the 256 bit key is calculated by applying the PBKDF2 key derivation function to the passphrase, using the SSID as the salt and 4096 iterations of HMAC-SHA1. WPA-Personal mode is available with both WPA and WPA2.

WPA-Enterprise:

Also referred to as *WPA-802.1X mode*, and sometimes just *WPA* (as opposed to WPA-PSK), this is designed for enterprise networks and requires a RADIUS authentication server. This requires a more complicated setup, but provides additional security (e.g. protection against dictionary attacks on short passwords). Various kinds of the Extensible Authentication Protocol (EAP) are used for authentication. WPA-Enterprise mode is available with both WPA and WPA2.

Wi-Fi Protected Setup (WPS):

This is an alternative authentication key distribution method intended to simplify and strengthen the process, but which, as widely implemented, creates a major security hole via WPS PIN recovery.

Encryption Protocol

TKIP (Temporal Key Integrity Protocol):

The RC4 stream cipher is used with a 128-bit per-packet key, meaning that it dynamically generates a new key for each packet. Used by WPA.

CCMP (*C*TR mode with *C*BC-*M*AC *P*rotocol):

The protocol used by WPA2, based on the Advanced Encryption Standard (AES) cipher along with strong message authenticity and integrity checking that is significantly stronger in protection for both privacy and integrity than the RC4-based TKIP used by WPA. Among informal names are "AES" and "AES-CCMP". According to the 802.11n specification, this encryption protocol must be used to achieve the fast 802.11n high bitrate schemes, though not all implementations enforce this. Otherwise, the data rate will not exceed 54 MBit/s.

EAP Extensions Under WPA and WPA2 Enterprise

Originally, only EAP-TLS (Extensible Authentication Protocol - Transport Layer Security) was certified by the Wi-Fi alliance. In April 2010, the Wi-Fi Alliance announced the inclusion of additional EAP types to its WPA- and WPA2- Enterprise certification programs. This was to ensure that WPA-Enterprise certified products can interoperate with one another.

As of 2010 the certification program includes the following EAP types:

- EAP-TLS (previously tested)
- EAP-TTLS/MSCHAPv2 (April 2005)
- PEAPv0/EAP-MSCHAPv2 (April 2005)
- PEAPv1/EAP-GTC (April 2005)
- PEAP-TLS
- EAP-SIM (April 2005)
- EAP-AKA (April 2009)
- EAP-FAST (April 2009)

802.1X clients and servers developed by specific firms may support other EAP types. This certification is an attempt for popular EAP types to interoperate; their failure to do so as of 2013 is one of the major issues preventing rollout of 802.1X on heterogeneous networks.

Commercial 802.1X servers include Microsoft Internet Authentication Service and Juniper Networks Steelbelted RADIUS as well as Aradial Radius server. FreeRADIUS is an open source 802.1X server.

Security Issues

Weak Password

Pre-shared key WPA and WPA2 remain vulnerable to password cracking attacks if users rely on a weak password or passphrase. To protect against a brute force attack, a truly random passphrase of 20 characters (selected from the set of 95 permitted characters) is probably sufficient.

Brute forcing of simple passwords can be attempted using the Aircrack Suite starting from the four-way authentication handshake exchanged during association or periodic re-authentication.

To further protect against intrusion, the network's SSID should not match any entry in the top 1,000 SSIDs as downloadable rainbow tables have been pre-generated for them and a multitude of common passwords.

WPA Packet Spoofing and Decryption

Mathy Vanhoef and Frank Piessens significantly improved upon the WPA-TKIP attacks of Erik Tews and Martin Beck. They demonstrated how to inject an arbitrary amount of packets, with each packet containing at most 112 bytes of payload. This was demonstrated by implementing a port scanner, which can be executed against any client using WPA-TKIP. Additionally they showed how to decrypt arbitrary packets sent to a client. They mentioned this can be used to hijack a TCP connection, allowing an attacker to inject malicious JavaScript when the victim visits a website. In contrast, the Beck-Tews attack could only decrypt short packets with mostly known content, such as ARP messages, and only allowed injection of 3 to 7 packets of at most 28 bytes. The Beck-Tews attack also requires Quality of Service (as defined in 802.11e) to be enabled, while the Vanhoef-Piessens attack does not. Neither attack leads to recovery of the shared session key between the client and Access Point. The authors say using a short rekeying interval can prevent some attacks but not all, and strongly recommend switching from TKIP to AES-based CCMP.

Halvorsen and others show how to modify the Beck-Tews attack to allow injection of 3 to 7 packets having a size of at most 596 bytes. The downside is that their attack requires substantially more time to execute: approximately 18 minutes and 25 seconds. In other work Vanhoef and Piessens showed that, when WPA is used to encrypt broadcast packets, their original attack can also be executed. This is an important extension, as substantially more networks use WPA to protect broadcast packets, than to protect unicast packets. The execution time of this attack is on average around 7 minutes, compared to the 14 minutes of the original Vanhoef-Piessens and Beck-Tews attack.

The vulnerabilities of TKIP are significant in that WPA-TKIP had been held to be an extremely safe combination; indeed, WPA-TKIP is still a configuration option upon a wide variety of wireless routing devices provided by many hardware vendors. A survey in 2013 showed that 71% still allow usage of TKIP, and 19% exclusively support TKIP.

WPS PIN Recovery

A more serious security flaw was revealed in December 2011 by Stefan Viehböck that affects wireless routers with the Wi-Fi Protected Setup (WPS) feature, regardless of which encryption method they use. Most recent models have this feature and enable it by default. Many consumer Wi-Fi device manufacturers had taken steps to eliminate the potential of weak passphrase choices by promoting alternative methods of automatically generating and distributing strong keys when users add a new wireless adapter or appliance to a network. These methods include pushing buttons on the devices or entering an 8-digit PIN.

The Wi-Fi Alliance standardized these methods as Wi-Fi Protected Setup; however the PIN feature as widely implemented introduced a major new security flaw. The flaw allows a remote attacker to recover the WPS PIN and, with it, the router's WPA/WPA2 password in a few hours. Users have been urged to turn off the WPS feature, although this may not be possible on some router models. Also note that the PIN is written on a label on most Wi-Fi routers with WPS, and cannot be changed if compromised.

MS-CHAPv20.25 in

Several weaknesses have been found in MS-CHAPv2, some of which severely reduce the complexity of brute-force attacks making them feasible with modern hardware. In 2012 the complexity of breaking MS-CHAPv2 was reduced to that of breaking a single DES key, work by Moxie Marlinspike and Marsh Ray. Moxie advised: "Enterprises who are depending on the mutual authentication properties of MS-CHAPv2 for connection to their WPA2 Radius servers should immediately start migrating to something else."

Hole196

Hole196 is a vulnerability in the WPA2 protocol that abuses the shared Group Temporal Key (GTK). It can be used to conduct man-in-the-middle and denial-of-service attacks. However, it assumes that the attacker is already authenticated against Access Point and thus in possession of the GTK.

Lack of Support of Forward Secrecy

WPA doesn't provide Forward secrecy, meaning that once pre shared key is compromised, an adverse person can decrypt all encrypted Wifi packets transmitted in the future and even past, which could be passively and silently collected by the attacker. This also means an attacker can silently capture and decrypt other's packets if a WPA-protected access point is provided free of charge at a public place like cafe because its password is usually shared to anyone with physical access without any special additional authorization. In other words, WPA only protects from attackers who don't have access to the password. Because of that, it's safer to use transport-layer security measure like TLS on top of that for the transfer of any sensitive data.

References

- Amoroso, Edward, "Intrusion Detection: An Introduction to Internet Surveillance, Correlation, Trace Back, Traps, and Response," Intrusion.Net Books, Sparta, New Jersey, 1999, ISBN 0-9666700-7-8
- Michael E. Whitman; Herbert J. Mattord (2009). Principles of Information Security. Cengage Learning EMEA. ISBN 978-1-4239-0177-8. Retrieved 25 June 2010
- Tim Boyles (2010). CCNA Security Study Guide: Exam 640-553. John Wiley and Sons. p. 249. ISBN 978-0-470-52767-2. Retrieved 29 June 2010
- Harold F. Tipton; Micki Krause (2007). Information Security Management Handbook. CRC Press. p. 1000. ISBN 978-1-4200-1358-0. Retrieved 29 June 2010
- John R. Vacca (2010). Managing Information Security. Syngress. p. 137. ISBN 978-1-59749-533-2. Retrieved 29 June 2010
- Robert C. Newman (19 February 2009). Computer Security: Protecting Digital Resources. Jones & Bartlett Learning. ISBN 978-0-7637-5994-0. Retrieved 25 June 2010
- Anderson, Ross (2001). Security Engineering: A Guide to Building Dependable Distributed Systems. New York: John Wiley & Sons. pp. 387–388. ISBN 978-0-471-38922-4
- Kohlenberg, Toby (Ed.), Alder, Raven, Carter, Dr. Everett F. (Skip), Jr., Esler, Joel., Foster, James C., Jonkman Marty, Raffael, and Poor, Mike, "Snort IDS and IPS Toolkit," Syngress, 2007, ISBN 978-1-59749-099-3

Permissions

All chapters in this book are published with permission under the Creative Commons Attribution Share Alike License or equivalent. Every chapter published in this book has been scrutinized by our experts. Their significance has been extensively debated. The topics covered herein carry significant information for a comprehensive understanding. They may even be implemented as practical applications or may be referred to as a beginning point for further studies.

We would like to thank the editorial team for lending their expertise to make the book truly unique. They have played a crucial role in the development of this book. Without their invaluable contributions this book wouldn't have been possible. They have made vital efforts to compile up to date information on the varied aspects of this subject to make this book a valuable addition to the collection of many professionals and students.

This book was conceptualized with the vision of imparting up-to-date and integrated information in this field. To ensure the same, a matchless editorial board was set up. Every individual on the board went through rigorous rounds of assessment to prove their worth. After which they invested a large part of their time researching and compiling the most relevant data for our readers.

The editorial board has been involved in producing this book since its inception. They have spent rigorous hours researching and exploring the diverse topics which have resulted in the successful publishing of this book. They have passed on their knowledge of decades through this book. To expedite this challenging task, the publisher supported the team at every step. A small team of assistant editors was also appointed to further simplify the editing procedure and attain best results for the readers.

Apart from the editorial board, the designing team has also invested a significant amount of their time in understanding the subject and creating the most relevant covers. They scrutinized every image to scout for the most suitable representation of the subject and create an appropriate cover for the book.

The publishing team has been an ardent support to the editorial, designing and production team. Their endless efforts to recruit the best for this project, has resulted in the accomplishment of this book. They are a veteran in the field of academics and their pool of knowledge is as vast as their experience in printing. Their expertise and guidance has proved useful at every step. Their uncompromising quality standards have made this book an exceptional effort. Their encouragement from time to time has been an inspiration for everyone.

The publisher and the editorial board hope that this book will prove to be a valuable piece of knowledge for students, practitioners and scholars across the globe.

Index

www.ingramcontent.com/pod-product-compliance
Lightning Source LLC
La Vergne TN
LVHW080906210926
848485LV00004B/602